Music, Time, and Its Other

Music, Time, and Its Other explores the relation between the enigmatic character of our temporal experiences and music's affective power. By taking account of competing concepts of time, Savage explains how music refigures dimensions of our experiences through staking out the borderlines between time and eternity. He examines a range of musical expressions that reply to the deficiency born from the difference between time and an order that exceeds or surpasses it and reveals how affective tonalities of works by Bach, Carolan, Debussy, Schoenberg, Messiaen, and Glass augment our understanding of our temporal condition. Reflections on the moods and feelings to which music gives voice counterpoint philosophical investigations into the relation between music's power to affect us and the force that the present has with respect to the initiatives we take. *Music, Time, and Its Other* thus sets out a new approach to music, aesthetics, politics, and the critical roles of judgment and imagination.

Roger W. H. Savage is a professor at the University of California, Los Angeles. His publications include *Hermeneutics and Music Criticism* and the edited volume *Paul Ricoeur in the Age of Hermeneutical Reason: Poetics, Praxis and Critique.* He was a Fulbright Scholar and a Moore Institute Visiting Fellow at the National University of Ireland, Galway.

Routledge Research in Music

For a full list of titles in this series, please visit www.routledge.com

6 **Masculinity in Opera**
Gender, History, & New Musicology
Edited by Philip Purvis

7 **Music in Films on the Middle Ages**
Authenticity vs. Fantasy
John Haines

8 **Popular Music in a Digital Music Economy**
Problems and Practices for a Service Industry
Tim J. Anderson

9 **Music, Performance, and the Realities of Film**
Shared Concert Experiences in Screen Fiction
Ben Winters

10 **The Modern Percussion Revolution**
Journeys of the Progressive Artist
Edited by Kevin Lewis and Gustavo Aguilar

11 **Preserving Popular Music Heritage**
Do-it-Yourself, Do-it-Together
Edited by Sarah Baker

12 **Opera in a Multicultural World**
Coloniality, Culture, Performance
Edited by Mary I. Ingraham, Joseph K. So, and Roy Moodley

13 **Current Directions in Ecomusicology**
Music, Culture, Nature
Edited by Aaron S. Allen and Kevin Dawe

14 **Burma, Kipling and Western Music**
The Riff from Mandalay
Andrew Selth

Music, Time, and Its Other

Aesthetic Reflections on Finitude, Temporality, and Alterity

Roger W. H. Savage

Routledge
Taylor & Francis Group
LONDON AND NEW YORK

First published 2018
by Routledge

2 Park Square, Milton Park, Abingdon, Oxfordshire OX14 4RN
52 Vanderbilt Avenue, New York, NY 10017

Routledge is an imprint of the Taylor & Francis Group, an informa business

First issued in paperback 2019

British Library Cataloguing-in-Publication Data
A catalogue record for this book is available from the British Library

Library of Congress Cataloging-in-Publication Data
Names: Savage, Roger W. H.
Title: Music, time, and its other: aesthetic reflections on finitude, temporality, and alterity / Roger W.H. Savage.
Description: Abingdon, Oxon: New York, NY: Routledge, 2018. | Series: Routledge research in music | Includes bibliographical references and index.
Identifiers: LCCN 2017016065 | ISBN 9781138679665 (hardback) | ISBN 9781315563473 (ebook)
Subjects: LCSH: Music—Philosophy and aesthetics. | Time—Philosophy. | Time in music.
Classification: LCC ML3800 .S176 2018 | DDC 781.2/2117—dc23
LC record available at https://lccn.loc.gov/2017016065

ISBN: 978-1-138-67966-5 (hbk)
ISBN: 978-0-367-23193-4 (pbk)

Typeset in Times New Roman
by codeMantra

Bach musicological font developed by © Yo Tomita

For Joanie

Contents

Music examples

Acknowledgments

My reasons for writing this book not only reflect my broader concerns with aesthetics, politics, and hermeneutical philosophy, but they also attest to the impact of innumerable discussions that have influenced my thinking. I recall a conversation with Michael Poellet many years ago that prompted me to consider how Bach's music evokes a sense of eternity. My interest in music's significance with respect to our temporal and historical condition is due in part to discussions that I have had, and continue to have, with Michael and with Wayne Turner.

I owe a special word of thanks to my sister, Joan Savage. Joanie has proven to be an indefatigable sounding board as I explored themes and ideas that ultimately found their way into this book. As a professional violinist and writer, she offered many insightful comments as well as provided thorough editorial remarks and suggestions on my manuscript.

My interactions with colleagues at the National University of Ireland, Galway, led me to think about aesthetics, politics, history, and memory in ways that I could not have anticipated. I want to thank Nessa Cronin, Verena Commins, Méabh Ní Fhuarthháin, and Tim Collins at the Centre for Irish Studies for their generosity in hosting me as a Fulbright Scholar, and for the lively exchanges that we had regarding politics, memory, history, and place. I want to thank the Fulbright Commission for the opportunity to conduct the research that appears in part in Chapter 5. I also want to thank Sam Murray for taking the time to tell me stories about Sailortown, Belfast, and for playing for me *The Suffering Ducks March*. Gary Hastings suggested that I contact Sam. If it were not for Gary, I might never have become aware of the story of the "suffering ducks."

Lillis Ó Laoire has been especially supportive in connection with my study on Irish lament in Chapter 5. I am grateful to Lillis for his extensive comments on drafts of this chapter. Any errors or omissions are my own. I want to thank Lillis for his assistance with the Irish texts that I have included, and for translating portions of the "Lament for Art Ó Laoghaire," Turlough Carolan's "The Parting of Companions," and Charles MacCabe's lament for Carolan. I also want to thank him for his friendship over the past twenty years.

Jennifer Fagre and Andrew Schnurr prepared the musical examples that appear in this book. I appreciate the care that they took in setting these examples for publication.

I also want to express my appreciation for the assistance provided by Mehrenegar Rostami. Negar's meticulous attention to reference, bibliographical, and formatting details has been invaluable in preparing the manuscript for publication.

As always, I am forever grateful to my wife, Pat, for her love and support, and for my daughters, Laura and Kristen, and son-in-law, Michael.

Finally, I wish to thank Universal Edition A.G. for permission to include an example from Arnold Schoenberg's *Erwartung*, and Hal Leonard for permission to include examples from Olivier Messiaen's *Quatuor pour la fin du Temps: pour violon, clarinette en si bémol, violoncelle et piano*, © Copyright 1942 – Éditions DURAND, Paris, France.

Introduction

Among all the diverse cultural practices that play a vital role in the life of a community, music holds a special place. Like myths, traditional stories, and contemporary popular culture, music affects our understandings and outlooks. The feelings and moods evoked by music open the world to us. On occasion, a world of sounds and sensations emanating from a musical performance lifts us out of the flow of ordinary events. In such moments, music's affective tonalities upend established conventions and habits of thought. Through exploring regions of feeling to which it gives voice, music attunes us to the world in new and sometimes unexpected ways by augmenting affective dimensions of our experiences.

The idea that music is most profoundly human when it stakes out the difference between time and its other sets music's power of expression in relief. This power springs from the way that music gives a figure to features and attributes exemplified through the manner in which music's constitutive elements cohere. In relating this power to the ways that music marks out the boundary posts between time and eternity, I am also attributing music's responses to the deficiency born from the difference between time and eternity to music's mimetic refiguration of time's ultimate unrepresentability. This deficiency is the emblem of the difference between the finite spans of our mortal lives and a plurality of expressions in which our experiences of time stand out against the backdrop of the mythic return to a time *in illo tempore*, the sempiternal revolution of the heavenly bodies, or the *nunc stans*. Music's evocation of feelings of "being beyond" or "out of time" attests to its power to allow time to be surpassed by its other. As such, multiple expressions in music of an order that exceeds our grasp explore regions of alterity in answer to the difference between time as we reckon with it in terms of our days, years, and lives and mythic time, cosmic time, or the *nunc stans*. We take hold of the way that this difference bears on our understanding of our temporal condition in rituals, cultural practices, and aesthetic expressions that sound out regions of experience only where our sense of time as stretching along or as relentlessly driving forward from the past into the future gives way to a renewed sense of time's ultimate unmasterability. Simulacra of a "time beyond time" are decidedly human ventures for searching out the limits of our temporally finite horizons. Accordingly, these fictions of "being beyond" or "out of time" reply to the enigma of time's inscrutable character through exploring the meaning of the difference between our temporal existence and the alterity of an order that exceeds or surpasses it.

Sound studies and research in anthropology and ethnomusicology amply attest that music has the capacity to construct unique worlds in which time is experienced in ways that are different from the quotidian order of the everyday succession of events. "Eternity" experiences that suspend or annihilate the regulated flow of time have a crucial place in my investigations into the ways in which music revalues time in this regard. However, these experiences (which in *Hermeneutics and Criticism* I identified as limit-experiences) do not exhaust the variety of modalities through which music replies to the difference between the being and the nonbeing of time. In the course of the reflections that I undertake in the coming chapters, I explore various ways in which music sublimates this difference between the being and the nonbeing of time in answer to time's deficiency vis-à-vis its other. As mortal beings, our cares, worries, and concerns are subject to the effects of time's passage; time's other is the cipher of the *nunc stans* over which time's corrosive effects have no power. It might therefore be tempting to credit the difference between time and eternity with a religious or theological significance. However, the fragility of our temporally finite existence (to which this difference attests) is not itself the sign of humanity's guilt or fault.[1] Rather, this difference is the emblem of the deficiency, or feeling of lack, that places its stamp on our experiences of time, which stands out against the backdrop of mythic and philosophical speculations on the being and the nonbeing of time.

This deficiency and music's responses to it constitute the central theme of this book. One of my first tasks in the next chapter will be to explain how a succession of sounds gives rise to expressive figures in response to the aporias that spring from the confrontation between the experience of time and time's representation as a series of "now" points (which I take up in Chapter 2). The feeling of motion commonly associated with a sense of forward-directed movement in traditional tonal works provides an initial point of access to music's exemplification of its affective attributes. Music's temporal extension is clearly a condition of its phenomenal sounding. Conversely, music's temporal features spring from the way that motives, themes, figures, and gestures cohere. Hence, my first task is to address the enigma of the feeling of motion that is produced by a sequence of tones that follow one another in succession but do not themselves move.

My account of how a succession of tones, harmonic sonorities, and the like give rise to a sense of movement paves the way to a broader reflection on time. Experiences of music in which time itself is seemingly obliterated or placed under erasure evoke a state or condition over which time's power holds no sway. These experiences thus evince the kind of deepening that sets time's other against our reckoning with the passage of our days and years.

Like the ritual return to a mythic time *in illo tempore*, these experiences renew time through taking the measure of the difference between time and its other. As I will argue, the pathos of the human condition inscribes itself in the dimensions of passivity born from the seeming collusion between our ultimate inability to master time and the alterity of an order presupposed by the attempt to constitute time in accordance with our capacities and powers. If I give special consideration to the absence of aseity in our temporally finite condition, it is in order to explore at length how music renews our inherence in the world in answer to time's ontological deficiency.

The hermeneutical approach that I take draws on the reflexive relation between music's mimetic character and philosophical investigations. Music not only brings to the fore phenomenological inquiries into the constitution of time through responding to them; by multiplying our experiences of time's deficiency vis-à-vis its other, music's affective resonances also sound out regions of feeling that impact the part we have in being. Music acquires its distinctively philosophical dimensions when, in answer to this deficiency between time and its other, music bears on the ways in which we interpret the meaning of the being and the nonbeing of time and the significance that it has with respect to the present. Just as philosophy sometimes demands to be read in terms of its musicality, so music sometimes affords the opportunity to think more in response to the experiences occasioned by it.[2] These experiences, I will argue in the last chapter, open the way to a renewed consideration of the relation between aesthetics, ethics, and politics. Consequently, music's power to break open the real, I will say, counterpoints the value of exemplary moral and political acts in refashioning or revolutionizing praxis.

Metaphysics and hermeneutics

The idea that music explores regions of feeling that are irreducible to ordinary sensibilities immediately risks being mistaken as invoking the specter of metaphysical transcendence. The notion that music—and especially purely instrumental music—takes flight in wordless expressions seems at once to authorize the conceit that music transcends the material order of social life. Furthermore, any suggestion that the experiences that we have of music resist their conceptual recuperation quickly stands accused of defending music's ineffability in order to safeguard music's metaphysical dignity. Isolated from the realities of social and political struggles, music, so the argument goes, is the last refuge of a bourgeois *mentalité* that sees its aesthetic education as its true spiritual calling.

Deconstructions of the claim that music is an aesthetically autonomous entity might at first glance appear to remedy the problem of music's metaphysical investiture. Music's supposed immunity to the exigencies, circumstances, and contingencies in which flesh-and-blood artists, practitioners, and listeners are caught up thus appears as the ideological apologetics for a discipline invested in preserving its canonic, "high-art" status. Yet by insisting that music is a worldly object subject to the same structures and forces that govern social life, critiques of music's chimerical metaphysical stature only reverse direction. When in the nineteenth century music's significance as a "language beyond language" assumed pride of place, the conceit of music's transcendent power became an article of faith. With the loss of this article of faith's credibility, a new orthodoxy supplanted an older (some might say modern) musicological dogma by discrediting the idea that music can be anything but worldly "through and through."[3] Lawrence Kramer consequently denounces the ekphrastic fear that shores up the disciplinary bulwark separating music's formal, aesthetic purity from verbal explanations of embodied constructions of sexuality, ethnicity, and subject identities.[4] Elevated to the level of a hermeneutical technique, *ekphrasis* justifies the practice of identifying music's discursive content through commenting critically on analogies and correlations between the structural processes and figures internal to music and these processes and figures' material, social correlates.

Music's elevation as an *opus absolutum et perfectum* clearly privileges the absence of definite meanings and extra-musical references as the preeminent feature of music's—and especially instrumental music's—aesthetic and metaphysical superiority. When in his review of Beethoven's *Fifth Symphony* E. T. A. Hoffman praises Beethoven's music, he expresses an aesthetic conviction that was gaining currency at the time. Beethoven's music, Hoffman exults, "reveals to man an unknown realm, a world quite separate from the outer sensual world surrounding him, a world in which he leaves behind all feelings circumscribed by intellect in order to embrace the inexpressible."[5] By setting "in motion the machinery of awe, of fear, of terror, of pain,"[6] Beethoven's music awakens the yearning for the infinite that Hoffman identified as the essence of romanticism. Like W. H. Wackenroder and Ludwig Tieck, Hoffman regarded absolute music—instrumental music that is free of programmatic associations and functions—as superior to vocal music.[7] Hence for Hoffman, Beethoven's vocal music was less successful than his symphonic works, since vocal music mired instrumental music's intimation of the absolute in concrete expressions of feelings as described by words.

The religion of art that subsequently came under attack with music's and art's exposure as ideologically complicit with dominant systems of thought took refuge in—and in a sense derived its justification from—absolute music's gnostic value. Drawing on Kant's thesis that "Fine art is the art of genius,"[8] German Romantic philosophers raised music and art above the prosaic order of material wants and needs by founding music's and art's aesthetic autonomy on the power of genius. Absolute music's metaphysical dignity traded on the romantic concept of aesthetic freedom, according to which a work is the product of a mind whose symbol-making powers are incomparably superior to the rational sobriety of society's practical concerns. By placing the mind's symbol-making activity at the heart of the aesthetics of genius, romantic philosophies of music and art capitalized on the felicitous connection between music's and art's claim to autonomy and this claim's metaphysical foundation. The happy "coincidence of sensible appearance and supersensible meaning"[9] in music and art rested on the expansion of a concept of the symbol (influenced by Goethe) into a then modern, universal aesthetic principle. In turn, this coincidence of phenomenal appearances and noumenal meanings founded the metaphysical link that supposedly leads "beyond the sensible to the divine."[10] Like the modern concept of the symbol, which, as Hans-Georg Gadamer points out, can be understood only in the light of its gnostic function, the claim that absolute music's wordless intimations of the infinite are premonitions of the absolute thus acquires its weight and force against the backdrop of the romantic cult of genius's metaphysical underpinnings.

Absolute music's apotheosis as *the* romantic art consequently ratifies the subjectivization of expression that finally liberated music from the material order of practical life. The ideal of somnambulistic production, which gives the spontaneous outflowing of the genius's creative powers pride of place, not only redoubles the poetic conceit that music's wordless artistry gives free reign to the subject's inner spirit. The metaphysical foundation of music's ineffable quality paradoxically also roots music's intimations of the infinite in the subject's creative powers. The advent of a culture devoted to the life of art would have been inconceivable apart from the fundamental achievement of a modern subjectivity that embraced this new mythology, which reached its apogee with the cult of *Bildung*'s consecration of music's aesthetically redemptive, transcendent power. Pierre Bourdieu rightly emphasizes that as "the most radical and most absolute form of the negation of the world, and especially the social world,"[11] music's flight from reality into metaphysical regions where, according to the romantic doctrine of a life educated to art, the human spirit is truly at home, is in actuality the sign of a life of ease. It is as though music's immaterial, ethereal substance is at once either the instrument of the artist's aesthetic divinations or the chimerical disguise of a cultivated culture's aesthetic tastes in struggles for domination and power.

Aligning music's transcendence of the real with the aesthetic mystification of music's ideological complicity with hegemonic constructions of gendered identities has become commonplace. Susan McClary, for example, claims to lay bare the sexual politics that she maintains are surreptitiously enacted by the traditional sonata form by deconstructing the "Master Narrative" of Absolute Music.[12] According to her, the tonal schema constitutes the framework of a sonata form movement in which the large-scale dissonance that she identifies with the "feminine" theme (which in the exposition is set against the "masculine" tonic theme) is subjugated for the sake of tonal closure. The narrative strategy that she adopts effectively equates the tonal structure of sonata form with a patriarchal order, in which the "feminine Other" is subdued or silenced in the interest of preserving the male protagonists' privileged position and authority. Similar to Hermann

Kretzschmar's musical hermeneutics, which he espoused almost a century before, McClary's narrativizing stance comes to the aid of an exegetical account of music's ostensibly material, social significance.

Mobilizing interpretive strategies that identify music's structural features and formal processes with socially constructed representations of gender, identity, and the like clearly has its advantages when it comes to dismantling the idea of absolute music. Yet hermeneutic techniques deployed in the interest of unmasking aesthetically dissimulated constructions that, like ideology, operate behind our backs, mask the blind spot of a musicological enterprise that otherwise aims at uncovering music's social, cultural, and political relevance. This enterprise purports to reclaim the cultural work performed by music. The notion that music affects us through conveying attitudes and conventions that inhere in its structures and processes clearly takes aim at metaphysical conceits. Supplanting sacrosanct aesthetic ideals with deconstructive critiques thus tears away the mask of music's metaphysical transcendence. Yet is music's efficacy vis-à-vis its cultural work limited only to representing (in the sense of redoubling or reproducing) existing norms, structures, and mores?

My disagreement with the way that hermeneutics has been enlisted in the service of sociological readings stems in large part from the occultation of music's worlding power. This worlding power springs from the way that a work exemplifies the affective tonalities to which it gives voice. A work's expression, I will explain in the next chapter, arises from the way that the work's constitutive elements cohere. The work's affective power, I will therefore say, is a function of the operation that draws a temporal configuration from the succession of tones, harmonic sonorities, timbral combinations, or acousmatic events. In the final chapter, I will explain how music's communicability rests on the listener's capacity to draw patterned figures from successively sounding events. Correlatively, music's expressive force springs from the work's exemplification of its affective attributes.[13] The work's configuration is accordingly the condition of the work's worlding power. Hence, a work is a sounding configuration capable of projecting a world that is replete with sensations, moods, and feelings. Consequently, the idea that a work is an *opus perfectum et absolutum* isolated from its sustaining life contexts is only the shadow of music's expressive vehemence.

In *Hermeneutics and Music Criticism*, I examine how the schema that legitimates music's claim to autonomy frames the riposte to music's aesthetic isolation from a supposedly lower order of reality. Kant's subjectivization of aesthetics proved to be the staging ground of art's—and ultimately music's—aesthetic dissociation from this practical order. The transcendental function that Kant ascribes to aesthetic judgment establishes the foundation for differentiating between art's aesthetic constitution and conceptual knowledge and truth. When in the nineteenth century music and art came to be regarded as the "practice of freedom,"[14] the separation of music and art from reality was complete. The schema of music's opposition to reality is in effect a result of a history in which Kant's justification of judgments of taste and the Romantic appropriation of his statement regarding genius and fine art play their parts. By reversing the terms of the original opposition in order to right the inversion that in Marx's optical vocabulary reverses the relation between the practical world and its ideological representations, deconstructive critiques risk preserving this schema, thereby perpetuating the effects of a history to which they are also ostensibly opposed.

One of my main concerns with dismantling the idea of a musical work as an *opus absolutum et perfectum* in this way is that it marginalizes music's mimetic significance. Music would be completely innocuous if not for its power to affect how we internalize our relation to ourselves, to others, and to the world. Apart from this power, music would have no ideologically deleterious effects. But why should music's efficacy be consigned only to fulfilling this role? One of the ironies of supplanting music's metaphysical stature with textual interpretations of its socially encoded meanings springs from the methodological restrictions imposed by various interpretive stances. The occultation of music's creative, productive potential thus paradoxically takes root in the strategic ambitions of disciplinary and cross-disciplinary practices aimed at overthrowing worn out dogma.

Setting music's creative potential against its dissimulating potential has the advantage of highlighting the relation between them. One of the conclusions that I draw in *Hermeneutics and Criticism* is that music's worlding power is the condition of the ideological role music plays in reproducing socially constructed representations of gender, sexuality, and identity. Cultural musicologists and theorists, ethnomusicologists, sound studies critics, and music philosophers have rightly rejected the idea that music operates independently of the circumstances in which it is created, performed, heard, and in which it circulates. Yet why, in view of my preceding comments, should we not also recognize that music evokes experiences that explore new regions of feeling? My hermeneutical approach differs from musicological studies, cognitive psychology, and even phenomenological investigations in this respect. Insofar as these different disciplinary investigations seek to explain and understand music's significance, I share their concern with the meaning of the experiences that we have in listening to sounds, ambient resonances of our surrounding environs, or musical works. At the same time, by placing the accent on music's mimetic significance, I also draw out the value that music has vis-à-vis its capacity to augment affective dimensions of our experiences in response to time's ultimately aporetic character.

The experiences occasioned by music are unique, especially insofar as music is a privileged mode for exploring regions of feeling in which time is surpassed by an order that in one way or another surges forth on the side of time's other. Emmanuel Levinas has alerted us to the challenge imposed by the failure of the hubris of the kind of thinking that dares to presume to master time. According to Levinas, time springs eternal as a transcendent event constituted by the irruption of a relationship with the Other that holds the horizon of the future in readiness.[15] The Other's exteriority is therefore also the limit of the subject's powers. The "out-of-time" experiences that music brings about similarly shatter the presumptions of a modern subjectivity that dares to proclaim itself as the master of history, time, and truth. Any analysis attentive to music's cultural significance could not fail to recognize that music's meaning and value spring from the solutions that music provides to the enigmas and aporias that provoke us to think more about the conditions that affect how we feel and act. In this sense, my undertaking in this book is a part of a larger hermeneutical inquiry into the fragility and vulnerability of human beings who are capable of acting but who also suffer due to circumstances and exigencies that are not of their own choosing.[16]

The solutions that musical works provide to the enigmas to which speculations on time give rise are accordingly a privileged point of access to the pathos that accompanies our power to act. Paul Ricoeur reminds us that the order of alterity that sets itself against the subject's (self-)mastery is one that corresponds to different experiences in which we become aware of how we are affected by our bodies, by others, and by our consciences.[17] Music's mimetic significance differs from that of the art of narrative in this regard. By privileging narrative as the mode through which time becomes human in response to the aporetics of temporality, Ricoeur places the accent on narrative's significance with respect to the field of action (*praxis*). Accordingly, music, he once suggested, is not mimetic. Setting music against the narrative emplotment of events, however, also brings to the fore the difference between music's and narrative's respective relations to the field of our everyday experiences. It is not surprising that playing music and telling stories have a place in virtually all the world's cultures. For where narrative imitates action, music gives voice to the pathos of our mortal condition. Like narrative, music's retreat from the real is the condition of music's capacity to explore new ways of relating to, and inhering in, the world. However, the nearer the narrative art draws to time's ultimate unrepresentability, the closer it approaches its own limits. The art of narrative encounters an internal limit when, in attempting to speak of time's inscrutable character, narrative exceeds itself to the point that it exhausts its own resources. Conversely, narrative reaches its external limit at the point where the "narrative genre overflows itself into other genres of discourse that, in their own ways, undertake to speak of time."[18] Beyond the limits of the narrative art, lyric poetry gives a voice to the grief that is ceaselessly born from the misery of life, which is subject to time's wearing effect. This voice, Ricoeur emphasizes, is also a song. Sometimes, we can hear in this song the mythopoetic echo of music's elegiac sublimation of the deficiency that strikes at the heart of our experiences of time.

Like ethnomusicologists, cultural musicologists, and sound studies scholars, the basic questions "What does music mean?" and "How and why is music meaningful?" figure prominently in my explorations in this book. I readily admit that these questions give rise to a plurality of methods, constructs, and intellectual convictions that resist any systematic integration. Furthermore, different disciplinary constructs and practices have their histories, which are punctuated by the ruptures, breaks, and conceptual shifts that reform or revolutionize them. The crisis of representation in anthropology in the 1980s, for example, raised a pressing political concern that motivated anthropologists, ethnomusicologists, and ethnographers to address the differential power relations between them and the individuals and groups with whom they worked.[19] This crisis not only brought into critical focus the fact that ethnographic representations are inflected by, and in a sense remain mired in, the ethnographer's own worldviews; the recognition of others' cultural realities also foregrounded the fact that since ideology is an inescapable condition of all social life, the hope of attaining an objective, epistemologically neutral standpoint is theoretically and practically impossible. This theoretical and practical impossibility is indicative of the hermeneutical situation in which we all find ourselves. Accordingly, the accounts we give and explanations we provide of music's significance and power rest on, and are rooted in, a history of prior questions, problems, and understandings out of which the particular questions, challenges, and problems that we take up arise.

My own investigations into the significance that music has with regard to our temporally finite condition take place on a number of fronts. In the reading guide that follows, I outline some of the connections that I explore between music, myth, religious experience, lament, and apocalyptic and eschatological representations of the end of time. The basis of my claim that music revalues time through refiguring time's ultimate unrepresentability rests on music's power to refashion our affective affinities with the world in response to the deficiency that is born from the difference between time and its other. I consequently emphasize this difference when drawing upon insights from musicology and sound studies. Furthermore, music's power to revalue time is a testament to the absence of aseity that places its stamp on the initiatives that we take. When in Chapters 6 and 7 I take up the question of the meaning of time in the context of aesthetic experience's lateral transposition onto the planes of ethics and politics, I will set music's mimetic refiguration of the difference between time and its other against the war against totality fueled by Jean-François Lyotard's enthusiasm for the postmodern sentiment of the sublime. Kiene Brillenburg Wurth rightly points out that romantic expressions of the sublime embrace the irony of a yearning for the infinite as its own (im)possibility.[20] Romantic expressions of the sublime are symptomatic of a modern subjectivity's fragility in this respect. Conversely, by voiding the imagination's schematizing power, postmodern inversions of feelings of terror and pleasure place the subject under erasure. The deficiency that places its stamp on our finite historical existence and on our experiences of time calls for a different response. Hence, I argue that the evocation of feelings and moods in individual works raises a claim to truth in answer to the difference between the being and the nonbeing of time.

Reading guide

Chapter 1 provides an account of music's expressive power that highlights the relation between music's temporal features and its affective attributes. Starting with the sensation of temporal movement, I explain how the feeling of motion is a function of the operation that transforms the temporal succession of a work's heterogonous elements into meaningful, patterned complexes. This operation is evident in the way that individual tones, which follow one another in succession, cohere as a melody, theme, or musical passage. A melody or theme accordingly takes shape through the way that a figure emerges from sequences of tones, harmonic sonorities, and timbral combinations. In attributing the affective properties exemplified by a theme or a passage to the theme or the passage's configuration, I consequently also relate the feelings evoked to a work's temporal constitution.

The hermeneutical approach that I set out in this first chapter has at least three advantages over traditional analytic orientations and interpretive stances. First, identifying music's affective properties with a work's temporal configuration overcomes the difficulty of basing the feeling of motion on the ordinary concept of time. According to this concept, time can be represented as a series of instantaneous "now" points along a line that extends from the past into the future. Alfred Schutz highlights the paradox that arises with regard to temporal movement along this line when the flight of an arrow, for example, is reduced to a series of such instantaneous points. Just as in this spatializing representation the arrow does not fly, conceiving tonal motion in terms of a succession of tones is trapped in the paradox that tones following one another in succession do not themselves move. Second, attributing tonal motion to the schematization of a work's temporally dynamic character highlights the connection between a work's affective features and its mimetic force. By showing how the work's retreat from the practical order of everyday experiences is the condition of its power to affect us, I relate music's efficacy to its mimetic redescription of affective dimensions of our experiences. The third advantage of the strategy I set out in the first chapter introduces the major theme of this book. By relating feelings and moods expressed by music to time's ontological deficiency, I lay the ground for exploring how music replies to the absence of aseity evinced by our temporally finite condition in response to the difference between time and eternity.

Chapter 2 situates my reflections on music's temporal and affective features within a philosophical consideration of two competing conceptions of time. Beginning with my discussion of the myth of Syrinx, I explain how the conflict between a cosmological concept of time (where time is related to movement) and phenomenological time (where time is experienced as distended across the present into the past and the future) gives rise to the enigma of time's ultimate unrepresentability. The myth of Syrinx recalls the origin of music's sublimating power in the call to the nymph who was transformed into the pipes that bear her name. Stripped of its etiological intention, this mythic tale reveals the significance of music's proximity to myth in response to time's deficiency in relation to its other. In contrast to the structural anthropologist's claim that music and myth both obliterate time, I attribute music's expressions of a "time beyond time" to music's mimetic ripostes to the aporia of time's inscrutability. Augustine's meditations on time and eternity underscore the enigma of the being and the nonbeing of time. Furthermore, they bear the trace of a Hebraic inspiration that, in contrast to a Greek one, speaks of time and its other in tones of lamentation and praise. Jonathan Kramer's typology of different species of musical time approximates the deepening of the experience of time that the phenomenology of time succeeds in bringing to light. Similarly, the price of the phenomenology of time's success is evident in the way that this phenomenology multiplies the aporias engendered by it. Music's mimetic and even mythopoetic power springs from the responses music provides to time's ultimately inscrutable character. Music, I therefore argue, revalues time by expressing feelings and moods in accordance with its evocations of an order beyond time, in answer to the difference between time and its other.

In Chapter 3, I explore how Johann Sebastian Bach's music marks out a region of religious aesthetics. By asking whether a feeling of height supersedes the distinction between Bach's sacred and secular works, I set the stage for a series of reflections in which I attribute the sense of eternity in Bach's music to its evocation of a presentiment of abiding grace. Unlike scholars who argue that Bach intended to neutralize or annihilate time, I identify the presentiment of eternity in Bach's music with its affirmation of the part that we have in being. Bach's declaration that music's sole purpose was for the glory of God alone (*Soli Deo Gloria*) raises the question as to whether his sacred works are more authentically devotional thanks to their settings of religious texts. Conversely, I also ask if his music's affective power makes listeners more receptive to the messages conveyed by these texts. Bach's music evinces the proximity of the feeling of height with the glory of the infinite. Accordingly, the sublimation in Bach's music of the deficiency evinced by the difference between time and its other is a wellspring of the feeling of joy.

The reflections that I undertake in Chapter 4 set Arnold Schoenberg's renunciation of tonal music's nomic principles against the temptation of a style of thinking that dares to master time by elevating itself to the level of the absolute. By drawing out the analogies between tonality's ontotheological resonances and a totalizing system of thought, I highlight the analogical relation between Schoenberg's negation of tonality and the loss of credibility of the Hegelian philosophy of history. In contrast to historicizing interpretations that regard Schoenberg's emancipation of dissonance as logically necessary, I argue that the initiative taken by Schoenberg in freeing dissonance from the requirement that it resolve bears the stamp of the crisis of a historical present that is caught between a past that has been surpassed and a future that recedes more quickly than it can be approached.

The aporia unleashed by refusing to force music into the mold of traditional tonal music's temporal coherence reaches its highest point of virulence in *Erwartung*'s expression of foreboding anxiety. *Erwartung*'s figuration of the eternity of a second that explodes in a thousand sensations is therefore more than just a modernist emblem of hope's despair. By relating the "bad infinity" of the negative counterimage of the *nunc stans* to our experience of time as a principle of dispersion and alienation, I argue that *Erwartung*'s expressive force sharpens the contrast between time and eternity. Hence, through a kind of recoil effect, Schoenberg's break with tonal strictures highlights the force that the present has as the time in which we take the initiative in response to the hubris of a system of thought that dares to grasp time as a whole in the eternal present.

The grief wrought by the death of a loved one or friend brings to the fore the contrast between the brevity of our mortal lives and the endless cycles of the seasons and years. In Chapter 5, I explore how lament sublimates the sadness that places its stamp on our finite existence through the work of mourning. Keening rites are proof of lament's role in giving voice to feelings of grief. In Ireland, the "merry wake" fulfilled its ritual function through restoring the community to health following the loss of one of its own. Laments consequently renew the time of the living through liberating the bereaved from the bondage of grief. Turlough Carolan's *The Parting of Companions* (*Scarúint na gCompánach*), which he composed upon receiving news of the death of his friend Charles MacCabe, is an enduring testament to the feeling of solicitude that makes such a parting a salutary tribulation. The bondage of grief, I argue moreover, has a political counterpart in the suffering endured by the Irish, in this case, under English rule. Eibhlín Dubh Ní Chonaill's lament for her murdered husband, Art Ó Laoghaire, is rife with protest against English injustice. The Jacobite tradition of prophetic poetry in Ireland counterpoints plaints against the pain of Ireland's captivity. The message of hope brought by the female figure who in this poetic genre personifies Ireland's maternal care has an eschatological ring that resonates with the political demand for freedom. This demand, which sounds the call that incites revolutionary fervor, echoes in the complaint against the political experience of exile in one's own land. Consequently, by suggesting that a deeper memory of exodus reverberates in the poem, "The Deserted Land," I set the stage for my discussion in the next chapter of the force of the present in making freedom a reality.

In Chapter 6, I take up the question as to how music revalues time in response to time's inscrutable character. The ritual fault that in mythic thought separates the sacred from the profane provides an initial index of the impact that music's explorations of the "end of time" have with respect to refiguring time's ultimate unrepresentability. The challenges and crises that we face in the light of our own historical experiences, however, bar any retreat to a mythic time *in illo tempore*. Consequently, the question that guides my reflections in this chapter (and which I take up at greater length in Chapter 7) highlights how various representations of the "end of time" stake out the difference between the being and the nonbeing of time. The allegorical gesture through which Time rejoins Eternity at the close of Olivier Messiaen's *Quatuor pour la fin du Temps* constitutes one such representation of the end of time; the traditional cadential closing's transvaluation in the finale of Philip Glass's *Einstein on the Beach* is another. The contra-finality of this transvalued closing's recurring repetition evinces the apocalyptic model's conversion into the sign of a time of crisis. The war that Lyotard insists must be waged against totality sets in relief a hermeneutical understanding of this time of crisis. In contrast to Kant's and Lyotard's estimations, I therefore argue that the sublime registers the sense of the disproportion that sets our experiences of time against the alterity of an order that surpasses or exceeds them. This disproportion is the sign of the absence of aseity that inscribes itself at the heart of our temporally finite condition. Music accordingly renews our inherence in the world in answer to the difference between time and the other of time through refiguring the boundaries between them.

My reflections in the final chapter on the place of imagination in music's exemplification of the feelings expressed by it converge with my critique of the connection between aesthetics and politics. Taking up a theme that I touched on in *Hermeneutics and Music Criticism*, I credit the distance that music and literary critics rightly maintain is the condition of music, literature, and art's capacity to subvert existing practices and habits of thought to the imagination's productive force. Accordingly, I argue that works refashion reality in accordance with the worlds projected by them through setting the imagination to work. Consequently, the experiences occasioned by a work can augment the practical field of our everyday experiences. The lateral transposition of aesthetic experience onto the planes of ethics and politics underscores how the work's power to affect our ways of feeling, thinking, and acting is the spring of the work's claim to truth. At the same time, music's power to affect us differs from moral and political injunctions issuing from exemplary acts.

The notion that music revalues time in response to the difference between time and its other is consequently central to my entire enterprise. The musical examples that I include in this book are related thematically to the trajectory of the reflections I outlined above. This trajectory is guided by my concern to explore how the affective tonalities to which music gives voice resonate with the temporal exigencies of our mortal condition. The musical examples that I discuss are consequently exemplars of the ways in which music bears witness to the absence of aseity born from the difference between the being and the nonbeing of time. The examples that I have chosen by no means exhaust the ways in which music replies to this difference. Yet, taken together, Debussy's *Syrinx*, Bach's *St. Matthew Passion*, Schoenberg's *Erwartung*, Carolan's *The Parting of Companions*, Messiaen's *Quatuor pour la fin du Temps*, and Glass's *Einstein on the Beach* and *Akhnaten* mark out the trajectory of a line of thought that examines the significance of music's mythic origins; its expressions of hope, anxiety, and grief; and its representations of the end of time. These works' expressions of joy, sorrow, and apocalyptic foreboding attest to the variety of ways in which music renews the meaning of time through sublimating the pathos at the heart of our mortal finitude. In the course of my reflections on works by Bach, Schoenberg, Carolan, Messiaen, and Glass, I therefore argue that music not only revalues time in staking out the borderlines between time and eternity, but through refiguring time's ultimate unrepresentability, it also opens the world to us and us to the world anew.

Notes

1 Placing time's deficiency vis-à-vis its other under the theological topos of a "fallen" humanity lends itself to a theodicy according to which the difference between the being and the nonbeing of time is the evil to be overcome.

2 See Babette Babich, "Mousike Techne: The Philosophical Practice of Music in Plato, Nietzsche, and Heidegger," in *Between Philosophy and Poetry: Writing Rhythm History*, ed. Massimo Verdicchio and Robert Burch (London: Continuum, 2002).

3 Lawrence Kramer, *Classical Music and Postmodern Knowledge* (Berkeley, CA: University of California Press, 1995), 17; see Lawrence Kramer, "Music Criticism and the Postmodernist Turn: In Contrary Motion with Gary Tomlinson," *Current Musicology* 53, (1993); Gary Tomlinson, "Musical Pasts and Postmodern Musicologies: A Response to Lawrence Kramer," *Current Musicology* 53 (1993); Susan McClary, *Feminine Endings: Music, Gender, and Sexuality* (Minnesota, MN: University of Minnesota Press, 1991).

4 Lawrence Kramer, *Musical Meaning: Toward a Critical History* (Berkeley, CA: University of California Press, 2002), 18. I have critiqued Kramer's treatment of the linguocentric predicament formulated by Charles Seeger in my *Hermeneutics and Music Criticism* (New York: Routledge, 2010).

5 Ernst Theodor Amadeus Hoffman, "Review of Beethoven's Fifth Symphony," in *E. T. A. Hoffman's Musical Writings: Kreisleriana, The Poet and the Composer, Music Criticism*, ed. David Charlton, trans. Martyn Clarke (Cambridge: Cambridge University Press, 1989), 236.

6 Hoffman, "Review," 238: Beethoven, Hoffman asserts,

> is accordingly a completely romantic composer, and is not this perhaps the reason why he has less success with vocal music, which excludes the character of indefinite longing, merely representing emotions defined by words as emotions experienced in the realm of the infinite?

7 See Carl Dahlhaus, *The Idea of Absolute Music*, trans. Roger Lustig (Chicago, IL: University of Chicago Press, 1989), 33.

8 Quoted in Hans-Georg Gadamer, *Truth and Method*, 2nd rev. ed., trans. Joel Weinsheimer and Donald G. Marshall (New York: Crossroad Publishing, 1989), 58; see Immanuel Kant, *Critique of Judgment*, trans. Werner S. Pluhar (Indianapolis, IN: Hackett Publishing Co., 1987), 175.

9 Gadamer, *Truth and Method*, 77–78.

10 Ibid., 73.

11 Pierre Bourdieu, *Distinction: A Social Critique of the Judgment of Taste*, trans. Richard Nice (Cambridge, MA: Harvard University Press, 1984), 19.

12 McClary, *Feminine Endings*; Susan McClary, "Paradigm Dissonances: Music Theory, Cultural Studies, Feminist Criticism," *Perspectives of New Music* 32,

no. 1 (1994); see Fred Everett Maus, "Music as Drama," *Music Theory Spectrum* 10 (1988); Anthony Newcomb, "Schumann and Late Eighteenth-Century Narrative Strategies," *19th Century Music* 11, no. 2 (1987).

13 In *Hermeneutics and Music Criticism*, I explain how individual works express feelings and moods by exemplifying them. Following Paul Ricoeur's analysis of Nelson Goodman's treatment of exemplification as the obverse of denotation, I argue that feelings and moods expressed by a work are coextensive with the work's internal arrangement. The metaphorical transfer of properties depicted by the work consequently overcomes the ruinous opposition between connotation and denotation. Accordingly, expression belongs to the order of representation inasmuch as the work exemplifies feelings and moods to which it gives voice (*Hermeneutics and Music Criticism*, 95–102). Roger Scruton questions the value of Nelson Goodman's theory, since for Scruton this theory founders on the idea that "some works of art possess the property of referring to an attribute that they also metaphorically possess" (Roger Scruton, *Art and Imagination: A Study in the Philosophy of Mind* [London: Methuen and Co., 1974], 223). Scruton accordingly maintains that the "aesthetic attitude is not one of discovery, and its end-point is not the knowledge of facts, whether about the work of art or about the world to which it 'refers'" (Scruton, *Art and Imagination*, 226), since for him a work's reference to specific feelings that it presents is necessarily detached from cognitive judgments of truth. See also Roger Scruton, *The Aesthetic Understanding: Essays in the Philosophy of Art and Culture* (Manchester: Carcanet Press, 1983).

14 Gadamer, *Truth and Method*, 82: Gadamer points out that "when Schiller proclaimed that art is the practice of freedom, he was referring more to Fichte than to Kant."

15 Emmanuel Levinas, *Time and the Other*, trans. Richard A. Cohen (Pittsburgh, PA: Duquesne University Press, 1987). For Levinas, the absence of the Other over which the subject could exert her hold marks out the horizon "where a personal life can be constituted in the heart of a transcendent event" (90).

16 See, for example, my essays "Fragile Identities, Capable Selves," *Études Ricoeuriennes/Ricoeur Studies* 4, no. 2 (2013) and "Judgment, Imagination and the Search for Justice," *Études Ricoeuriennes/Ricoeur Studies* 6, no. 2 (2015).

17 Paul Ricoeur, *Oneself as Another*, trans. Kathleen Blamey (Chicago, IL: University of Chicago Press, 1992), 318. Ricoeur identifies three modalities of passivity that correspond to different centers of otherness. First, the experience of one's body ties the body's corporeal passivity to the mode of our incarnation in the world. For an ontology of the flesh, the dissociation of the notion of existence from that of substance frees the lived body from objectifying representations, so that one's own body becomes the organ and support of the initiatives that one takes. The dialectic of *praxis* (action) and *pathos* (affect) takes root in this fundamental register. For Ricoeur, the "certainty of the acting self and its contrary … [, which] is also its complement, corporal passivity" (321), gives rise to this dialectic. The second modality arises in connection with the confrontation with others. Moral and legal injunctions that hold an agent accountable for her actions underscore how the power exerted by one person can adversely affect another. In contrast to the ego that posits itself, the self recognizes that it is one among others. The third center of otherness is that of conscience. As the voice of the Other, conscience bears witness to the obligation that places one in another's debt. The appeal to conscience, replete in the metaphor of the silent, interior voice that addresses us, testifies to the vertical dimension of this call. Through heeding this voice, we yield to the injunction or command that issues from this call.

18 Paul Ricoeur, *Time and Narrative*, vol. 3, trans. Kathleen Blamey and David Pellauer (Chicago, IL: University of Chicago Press, 1988), 271.
19 See George E. Marcus and Michael M. J. Fischer, *Anthropology as Cultural Critique: An Experimental Moment in the Human Sciences* (Chicago, IL: University of Chicago Press, 1986).
20 Kiene Brillenburg Wurth, *Musically Sublime: Indeterminacy, Infinity, Irresolvability* (New York: Fordham University Press, 2009).

1 Music and time

Music, it has often been said, is the art of time. By ordering tones, rhythmic impulses, harmonies, and timbral colors, each work creates its own unique world. Within each world, the temporal succession of heterogeneous elements evokes feelings and moods to which the work gives voice. The power of this art of time thus takes root in the way that the successively sounding elements cohere. Hence, one of the first challenges for any inquiry into music's expressive value is to account for the link between music's temporal features and its affective attributes. In the course of this chapter, I will explain how music's expressive character and temporal features are a function of the operation that draws together successive tones, harmonic sonorities, and rhythmic impulses. This operation gives rise to music's expressive attributes; this synthetic operation, I will therefore ultimately say, produces the figures that possess these attributes through schematizing them.

The concept of time that has tended to dominate the way that we think about music's temporal character presents a number of difficulties in this respect. According to the widely accepted concept, time can be represented as an infinite series of instantaneous "now" points stretched out along a line. The first question we should therefore ask is how an ordered progression of individual tones gives rise to figures and gestures that possess their own distinctive characteristics. This question, which I will take up at some length, marks the entry into my broader hermeneutical investigation into music's significance vis-à-vis the difference between time and its other.

How, we could additionally ask, could a musical passage convey the sense of its progression if time were nothing more than an abstract measure of passing events? Correlatively, how could we apprehend a work's expressive features if music consisted only of a series of isolated elements that followed one another in succession? The ordinary concept of time, the rectilinear order of which is evident in the metaphor of the arrow that travels in only one direction (from the past to the future), leads to identifying music's temporal features with time's inexorable advance. Yet, as I will show, representations of temporal movement in music and even its opposite—temporal stasis—are functions of the operation that transforms successively sounding events into temporally meaningful expressions.

Form in motion

The phenomenon of tonal motion is the gateway to my broader study in this chapter on the interrelationships between music's temporal character, its expressive force, and its mimetic power. The question "How does music move in time?" provides a point of access to the initial difficulty of disentangling the feeling of motion from the ordinary representation of time. As an art of time, music takes shape through its unfolding movement. This movement therefore seems at first to consist in the succession of sounding events. Temporal succession is admittedly one aspect of a work's temporal formation. Yet, in listening to a tune, we hear the melody through grasping together the individual tones. The idea that one tone leads to another reveres the relation between the melody's temporal formation and the way that the succession of individual tones contributes to it. The feeling of motion that seems logically to inhere in the structured ordering of tones is thus a function of the process through which we draw the figure of the melody from a series of tones that follow one another in succession.

By asking whether music situates itself "in time" or has "time in itself,"[1] Carl Dahlhaus highlights the difficulty that arises when attempting to place the sensation of movement in music under the rule of the ordinary representation of time. Dahlhaus points out that if time is the medium for musical processes that take place within it, then time belongs to the order of motion and change (which I will explain, in the next chapter, is the order of cosmological time). If, on the other hand, time is itself "an occurrence, which advances from the past into the present or approaches the present from the future,"[2] the experience of time is one for which the phenomenology of time attempts to account (as I will also explain more fully in the next chapter). How, we might wonder, can we measure time's passage without referring to an order of time that exists independently of our experiences of it? Dahlhaus rightly insists that, like claims that music is a "shaping of time" (*Zeitgestalt*), claims that music is a "shaped time" (*gestaltete Zeit*) fail to resolve the fundamental question concerning a work's inner unity, on which the feeling of motion ultimately depends. If no one seems to doubt that "music is sounding motion,"[3] it is by no means obvious how this sounding motion is grasped as such in view of these two competing concepts of cosmological and phenomenological time. To what degree, Dahlhaus therefore asks, is the work's inner unity experienced passively, and to what extent is this unity the result of the spontaneous activity on the part of the listener in assembling and comparing constitutive parts in the process of listening to the work? By shifting the focus from the experience of time's passage to the inner unity constituted by the succession of a work's various component elements, he subsequently ties the one to the other by outlining their mutual implication.

Dahlhaus's proposed solution to the enigma of tonal motion seems at first to be ingenious. Drawing upon Henri Bergson's concepts of *temps durée* and *temps espace*, Dahlhaus sets the feeling of motion associated with a work's formative processes against some "fixed" temporal backdrop. The *temps durée*, according to which time is experienced as passing, and the *temps espace*, where time is imagined as being extended spatially, work together to produce a sense of movement and change. Dahlhaus recommends abstracting the *temps espace* from the *temps durée*, since for him the "stretchings and shortenings of experienced time can be felt only against a backdrop of spatial time."[4] Held fast through its "spatial" retention, a phrase of music can therefore be "imagined as lying in a *temps espace* resulting from its original *temps durée*."[5] The *temps espace* and the *temps durée* operate in consort to constitute the temporal frame in which the sensation of movement arises. The *temps durée* is accordingly the generative source of a work's spatio-temporal framework. Having first been heard in terms of its own temporal flow, each musical phrase passes into the temporal space that it progressively builds up. In this way, the *temps durée* engenders the framework within which a work's temporal hesitations and now furtive, now strident advances flesh out the structure of the flow of time.

However convincing it might appear to be, Dahlhaus's account gives rise to a number of paradoxes. The first paradox arises from the way that the feeling of motion (or of a forward-directed progression in the case of traditional tonal music) springs from the temporal unity of a sequence of tones that do not move but instead follow one another in succession. Musical gestalts, which consist of figures comprised by groupings of tones, are indicative of the way that the synthetic operation that transforms sequences of tones into expressive melodic figures resolves this paradox. The teleological thrust so widely attributed to tonal music is predicated on the way that successive thematic and harmonic elements contribute to a passage's or a work's formation. Even the cyclic recurrence of repeating patterns in minimalist repetitive music exploits the recurring succession of these cyclically repeating patterns to annihilate any sense of forward-directed movement. In this respect, music's temporal character distinguishes it from a sculpture, a painting, or even a text, the meaning of which is fixed by writing.[6] Not only is music's performative expression irreducible to a series of pinpointed instances, but the feeling of forward-directed motion or its opposite (temporal stasis) are both effects of the way that figures drawn from tones, harmonies, and timbral combinations that sound in succession cohere. Dahlhaus's solution to the paradox that sounds that follow one another in succession elicit a feeling of motion might suggest that a dynamic relation exists between the succession of sounding elements and a theme's, passage's, or work's temporal configuration as a whole. But his account of the relation between a *temps durée* and a *temps espace* in the end overlooks the operation from which expressive figures, such as motives and themes, spring.

Second, it is difficult to see how Dahlhaus's deployment of the concepts of a *temps durée* and a *temps espace* could account for the initial sense of movement of a work's opening phrase. How could we hear the progression of the first tone to the next as animating the opening phrase's first impulses when there is not yet any frame of reference built up by some prior phrase's retention? How, in other words, could the initial sense of movement be set against some fixed referent if there is no musical unit that is held fast in the imagination? This second paradox follows from the first. The idea that music is a *poiesis a se*, and thus creates its own time through its self-shaping movement, underscores the problem that arises in setting the feeling of motion against some fixed temporal frame.[7] Accordingly, this second paradox brings us closer to the problem at hand by pointing up the difficulty, if not the impossibility, of accounting for music's temporal character based only on the succession of sounding tones.

The third paradox raises the question of this temporal character's animating force. Dahlhaus's allusion to the stretching and shortening of time highlights the elastic qualities of the figures that evoke perceptions of time as slowing down, broadening out, or speeding up. Consequently, these qualities are psychological rather than physical. Time can feel stretched out or shortened depending on the character of a musical phrase, passage, or movement. A series of long, sustained tones supported by a languid harmonic rhythm might create the feeling that the flow of time is being temporally retarded. In contrast, an accelerated harmonic rhythm where the succession of harmonies quickens could elicit a sense of temporal compression. We might be tempted to regard a musical phrase's or passage's temporal flow in terms of its spatio-temporal displacement or movement. Yet what fixed point of reference is there for registering this displacement or movement? Dahlhaus admits that the "ideal of tonal space represents an abstraction from the phenomenon of musical motion."[8] Accordingly, like this movement itself, the force that animates it is a function of the process through which expressive figures and gestures cohere.

Referring tonal motion to abstract representations of tonal space consequently voids the dynamic relation between successively sounding tones and the expression of the figures in which senses of motion or stasis crystallize. Dahlhaus hoped to rescue his solution from the abstraction on which it depends by singling out rhythm as the essential component of the feeling of tonal motion. Rhythm for him is accordingly more primordial with respect to the manifestation of the distance or space separating successions of tones than are chords (harmonic sonorities). Yet, just as a melody takes shape through the way that tones follow one another in succession, we grasp a rhythm through apprehending a pattern drawn from a structured succession of separate impulses. Despite his effort to differentiate between the rectilinear and vertical dimensions of tonal space, Dahlhaus did not account for the dynamic relation between successive rhythmic impulses or harmonic sonorities and their temporal configuration. Nor could he account for this dynamic relation's inspiriting power. This inspiriting power reveals itself in the movement through which a work takes shape. Hence, as I will argue, the spirit of a work is the spring of its worlding power—that is, its power to augment our affective affinities with the world through evoking moods and feelings. The force that we might otherwise attribute to the movement of tones, the succession of harmonies, and rhythmic impulses thus proves to be a function of the process through which melodic figures and rhythmic structures crystallize.

Tones and impulses can obviously be plotted out, and the distances between them can be measured according to some standardized scale (quarter notes, eighth notes, seconds, or even microseconds in the case of "participatory discrepancies"[9] in performance). The notion that time consists of a series of instantaneous points projected along a spatial axis in fact lends itself to these types of measurements. Metaphorical allusions to melodic movements and harmonic progressions reinforce the idea that music's temporal features are amenable to ordinary representations of time. The idea that rhythmic movement can be set against a spatial referent redoubles the common-sense notion that quasi-physical accounts of music's temporal movement suffice. In view of the doxa that tonal motion is a function of the way that the pattern traced out by successive tones moves through or across a tonal space, the task at hand is to show how tonal forms in motion give rise to music's expressive power.

Eduard Hanslick, who has been widely regarded as the progenitor of formalist musical aesthetics, is an unlikely ally in this regard. Hanslick's thesis—that music's sole content consists of "audible tonal forms"[10]—aimed to discredit romantic conceits regarding absolute music's ineffable essence. At the beginning of his treatise *On the Musically Beautiful*, he admonishes serious students of musical aesthetics to reject the dubious practice of subjecting music to poeticizing interpretations of feelings. By setting musical aesthetics on a scientific footing, Hanslick intended to distance himself from romantic aesthetics and its outworn doctrine of expression. For him, feelings could no more constitute the work's idea than this idea could "attire... itself in the earthly body of physical sound in order that it may walk on earth here below as a musical artwork."[11] In order to rescue musical ideas from their metaphysical trappings, he renounced the transcendental connection between phenomenal appearances and supersensible meanings. Accordingly, he sought to rid aesthetics of this dubious doctrine by identifying music's content with the internal workings of its tonal forms.

By insisting that music's sense and logic is comparable to that of a logical proposition, Hanslick lays the ground for formalism's incursion into musicological discourses. At the same time, he opens the door to a hermeneutic phenomenology of music by reserving a place for the imagination's role in schematizing the operation that draws a musical figure from the temporal succession of sounding events. According to Hanslick, music possesses sense and logic inasmuch as a sequence of tones creates the impression of comprising a complete statement. Like a complete statement, such a sequence of tones is comprehensible only in that the individual tones form a meaningful unit. Hanslick stresses that the "layout of notation in the score as it presents itself to the eye"[12] is an empty abstraction. For only "that which is actually sounding"[13] expresses the work's ideas in a way that is consistent with their purely musical determination. Hence, the rational coherence that he identifies with the structural organization of a group of tones should in the first instance be attributed to this group of tones' temporal configuration. Sense and logic are accordingly a function of the schema that results from the process of drawing a patterned figure from successively sounding tones. This temporal schema constitutes the matrix from which the "rational" network of tonal relations and progressions is derived (see Chapter 7). Sense and logic in music are therefore inversely comparable to that of a logical proposition, since in the case of this temporal schema the "logical" content of a phrase, passage, movement, or work is abstracted from the sense of the fit evinced by the manner in which the individual tones cohere.

Hanslick's oft-cited statement, that music's sole content is "tonally moving forms,"[14] highlights the barely noticeable tension between his formalist leanings and his proto-phenomenological inclinations. His allegiance to positivist currents in the human sciences compelled him to seek out a rational foundation for beauty in music. Extracting music's logic from a temporal schema lent itself to the objective style of inquiry that he favored. By predicating the "gratifying reasonableness"[15] of music's structures on natural laws that ostensibly governed the human mind as well as sound, Hanslick presumed to legitimate the scientific standpoint that he adopted. In his view, these natural laws resided "instinctively in every cultivated ear, which accordingly perceives the organic rational coherence of a group of tones, or its absurdity and unnaturalness, by mere contemplation, with no concept as its criterion or *tertium comparationis*."[16] From this standpoint, the mysterious affinities and bonds that some romantic philosophers and critics identified as the source of ineffable feelings are governed by laws that are not only discoverable but are also the ground of musical beauty.

If, on the one hand, Hanslick sets out the framework for a formalist approach that concentrates on a work's internal structures and systems of organization to the exclusion of a more comprehensive understanding of music's contextual meaning or of the experiences that music occasions, on the other hand, his oblique allusions to music's temporal character mark out the beginnings of a phenomenological sketch of music's affective power. However much his statement, "Music is play,"[17] might seem to echo Hans-Georg Nägeli's theory regarding the play of musical forms, the implicit emphasis that Hanslick places on the temporal character of music's content is more closely related to Hans-Georg Gadamer's phenomenological analysis of the mode of being of the work of art. Phenomenologically, the work of art occasions an experience in accordance with the way that the work unfolds. If, as Hanslick maintains, the content of a work can be grasped only musically and not graphically, it is because the work itself acquires its force and significance through the play of tonal forms that brings the work to presentation. The world projected by the work is therefore a world that the work presents.[18] Hence, this world exists only in its temporal realization. Accordingly, the play of tonal forms is the spring of a work's worlding power. This phenomenological insight remains latent in Hanslick's treatise. Yet the fact that he refuses to identify music's essential content with the notational layout of the score cannot be overlooked. However much Hanslick laid the ground for formalism's incursion into music theory and analysis, there is no mistaking the fundamentally temporal aspect of his supposition that the content of music consists of sounding forms in motion.

The attacks that have been mounted against Hanslick's assertion that music is a self-sufficient art have either dismissed too hastily or entirely overlooked the role of imagination with respect to music's expressive power. Hanslick's references to imagination might be taken as a sign of a link with romantic sensibilities that he wanted to sever. By the same token, his adoption of the concept of absolute music preserves the ideal of aesthetic autonomy consecrated by the romantic religion of art. To be sure, Hanslick maintains that music's ideal content can be found only in a work's tone-structure. As such, the beauty that he identifies with the play of tonal forms is self-subsistent. However, there is no reason to conclude that by closing itself off from the practical order of our everyday experiences, a work sacrifices its claim to meaning or truth. On the contrary, the work's retreat from the real is the condition of its capacity to come bursting into the midst of our everyday experiences, as I will argue more fully below. The suspicion that a work's *epoché* of the real is the handmaiden of metaphysical conceits thus fuels the prejudice against a work's power to refashion dimensions of our experience, which in the Introduction I credited in part to the reversal that supplants music's metaphysical stature with critiques of music's ideological dissimulations of material constructions of gender, sexuality, and identity.

By attributing tonal connections apprehended by the listener to the "spontaneous activity of the imagination,"[19] Hanslick not only opens the door to this hermeneutical understanding of music's efficacy, if only by a crack, but he also invites a hermeneutical riposte to the romantic fallacy that a work embodies its composer's emotions, feelings, and thoughts. The principle of interpretive fidelity enshrined in the concept of *Werktreue* trades on the notion that a work's ideal performance fulfills its composer's intentions.[20] Conversely, music's worlding power springs from the way that its constitutive elements, which sound in succession, cohere. In view of the spontaneous activity that Hanslick attributes to imagination, we could not fail to ask how "each individual factor in a musical passage necessarily contributes to its taking on its own unique ideal expression."[21] Following Hanslick, we could also say that music has an effect on the listener only in terms of its (self-)expression. Hanslick stresses that the beauty that inheres in a self-subsistent theme makes itself known in a way that "permits no other explanation than the inner appropriateness of the phenomenon."[22] For him, this inner appropriateness constitutes the justification for extracting the logic that for music theorists and analysts is the ground of a work's rational structure and coherence. Whatever metaphysical trace of the cult of genius might haunt Hanslick's reference to imagination, his insight that a passage's or a work's distinctive character springs from, and hence is attributable to, the way that individual factors contribute to its overall configuration owes its force to the synthetic operation that draws a figure from a series of tones. Hanslick, however, could not cross this crucial threshold. Accordingly, he was unable to answer the hermeneutical question that critics of musical formalism by and large also pass over in silence: How is the work's inner appropriateness—and hence its expression—communicable through the play of tonal forms?

Hanslick and Hegel

The sense of inner appropriateness that Hanslick inadvertently placed at the center of his aesthetics marks a critical turning point in my discussion of music's temporal features and affective attributes. The emphasis that he put on the fit of individual factors that contribute to a passage's expression invites us to consider how the way that these constitutive components cohere manifests the ethos or spirit emanating from a musical passage's temporal unfoldment. By setting the imagination's productive role against the cult of genius and the cult of *Bildung* for which it served as midwife, I am admittedly extending Hanslick's interpretation beyond what he intended. The fit of a musical passage, which constitutes the matrix from which the logic of the passage is subsequently derived, attests to the imagination's schematizing power. Hanslick was not wrong to insist that the idea expressed by a theme is specifically musical. All the same, musical formalism's institution as a disciplinary mainstay eclipsed Hanslick's more fundamental insight into the fact that a passage or work presents its idea in accordance with the way that the passage or work as a whole unfolds. There is no reason to think that Hanslick should have accounted for the musical logic on which he based his scientific program in terms of the fit of heterogeneous elements produced by the way that these elements cohere. Similarly, there is no reason to think that he should have identified music's temporal figuration of its ideas with its affective power, even though his remarks to which I referred above provide a strong indication of this connection between the sense of a passage's inner appropriateness and its expressive force. Hanslick's concerted effort to rid aesthetics of the nineteenth-century's metaphysics of feeling undoubtedly blocks the way to these hermeneutical insights. Nevertheless, by giving the play of tonal forms a central place in his theory of the beauty in music, Hanslick opens a path to a hermeneutical account of music's expressive vehemence from within an otherwise formalist foundation.

Hanslick's antipathy toward Hegel's aesthetics sets in relief the connection between the ethos that emanates from a work and the play of tonal forms. Like Hegel, Hanslick identifies music's substance with an innervating spiritual force. For Hanslick, "thoughts and feelings [that] run like blood in the arteries of the harmonious body of beautiful sounds"[23] animate the spontaneous creations of a mind that works in, and expresses itself through, the medium of sound. According to Hanslick, the theme thus functions as the bud of a blossoming whole, as though everything in the work were the natural continuation and consequence of an initial musical idea. The contours of the work's inner processes are proof that a work's form is the product of a mind or spirit (*Geist*) that gives shape to itself within. The sense of the work's rightness, which Hanslick insists must be subjected to analysis in order to demonstrate how the work's structural integrity develops axiomatically from its initial theme, therefore harbors the trace of music's metaphysical ideal.

Although Hanslick excised the passages alluding to music's metaphysical value from his 1854 edition of *The Beautiful in Music*, the gnostic overtones of his statement that the "spiritual energy and distinctiveness of each composer's imagination make their mark upon the product as character"[24] betrays a continuing allegiance to romantic ideals. Influenced by Wilhelm von Humboldt's philosophy of language, which introduced the idea of "inner form" to nineteenth-century thought, Hanslick inverts Hegel's thesis that the phenomenal appearance of a work's animating spirit constitutes its true substance and content.[25] The crisis of confidence that Kevin Karnes notes subsequently led Hanslick to reject his positivist approach bears out Hanslick's allegiance to romantic doctrines. Motivated by a "faith in the Hegelian maxim that works of art are phenomenal manifestations of the human spirit in its evolution through time,"[26] Hanslick turned from his scientific program to music criticism and cultural history.

Hegel's understanding of music as the "ensouling of matter"[27] offers a more pointed insight into the romantic notion that the spirit of a work manifests itself through the form to which it gives shape and life. For Hegel, music's "expression must be the direct communication of a living individual who has put into it the entirety of his own inner life."[28] The revelation of Spirit's inner subjective unity is therefore in each case the work of a "specifically gifted [individual's] spirit."[29] Music reveals the inner life of this spirit as it presents itself in the movement of sound. Hence, the "chief task of music consists in making resound … the manner in which the inmost self is moved to the depths of its personality and conscious soul."[30] In the context of Hegel's aesthetics, music is capable only of expressing the element of feeling in which the innermost life of the soul, freed from all external objectivity, recognizes itself. Hence, music's "content is what is subjective in itself."[31]

Sound is therefore the ideal medium for bringing Spirit's inner subjectivity to presentation (*Darstellung*): for sound cancels all spatial externality. The phenomenal externality of sound is one that "in its coming-to-be is annihilated again by its very existence."[32] Hence, sound sublates Spirit's external objectification in sculpture and painting by vanishing with its passing. The state of inwardness that Hegel attributes to music is thus the only "form in which music can conceive its subject-matter."[33] Through adopting "everything which can enter the inner life as such and which above all can be clothed in the form of feeling,"[34] music has the expression of this inner life as its sole content.

Hegel's remarkable passage concerning this inner life's subjective unity merits citing at length:

> The inner life in virtue of its subjective unity is the active negation of accidental juxtapositions in space, and therefore a negative unity. But at first this self-identity remains wholly abstract and empty and it consists only in making *itself* its object and yet in cancelling this objectivity (itself only ideal and identical with what the self is) in order to make *itself* in this way a subjective unity. The similarly ideal negative activity in its sphere of externality is *time*. For (i) it extinguishes the accidental juxtaposition of things in space and draws their continuity together into a point of time, into a "now". But (ii) the point of time proves at once to be its own negation, since, as soon as *this* "now" is, it supersedes itself by passing into another "now" and therefore reveals its negative activity. (iii) On account of externality, the element in which time moves, no truly *subjective* unity is established between the first point of time and the second by which it has been superseded; on the contrary, the "now" still remains always the *same* in its alternation; for each point of time is a ""now" just as little distinguishable from the other, regarded as merely a point of time, as the abstract self is from the object in which it cancels itself, and since this object is only the empty self itself, in which it closes with itself.[35]

The negative unity that Hegel identifies with the phenomenal presentation of spirit's subjectively inner life anticipates the aporetic nature of the experience of time, which the phenomenology of time expressly brings to light (as I will elaborate in the next chapter). By canceling the external order of time (according to which time can be measured by reference to a clock, for example), the spirit animating a work presents itself as its own subject through extinguishing spatial juxtapositions. This spirit's inner subjective unity is something other than a series of instantaneous "now" points that succeed one another in succession. Through negating both the spatial plane of the painted canvas and its own external materiality, music accordingly manifests the inwardly self-conscious life of spirit in its subjective freedom. For Hegel, hearing is a more ideal sense in that, in listening to music, one apprehends "the first and more ideal breath of the soul."[36] Music releases the free expression of beauty in art from its entanglement in external matter. Consequently, music—which "has as its *content* the inner life itself, the inner sense of feeling"[37]—takes Spirit's self-actualization in its purely inward form as its proper subject-matter.

Since, for Hegel, Spirit alone is "true, comprehending everything in itself, ... everything beautiful is truly beautiful only in sharing in this sphere [of Spirit's self-comprehension] and generated by it."[38] Music's "*form*, sound, which in an art that least of all proceeds to spatial figures, is purely evanescent in its perceptible existence."[39] The state of pure inwardness that Hegel identifies with music's essential content represents an important stage in Spirit's advance toward absolute self-knowing. In Hegel's system, only poetry occupies a higher place than music in the pantheon of Romantic arts, which also includes painting and sculpture. However, music's withdrawal from the external order of spatial motion and movement also offers a glimpse of the difficulty raised by the supposed cancellation of time as external to spirit's inner subjective unity. For a philosophy that dares to elevate itself to the level of the absolute, music is "concerned only with the completely indeterminate movement of the inner spirit and with sounds *as if* they were feeling without thought."[40] Yet as we will see, the enigmas of time, the difference between time and its other, and the deficiency that sets our experiences of time against the Hegelian temptation to grasp time as a whole in an eternal present set the way we think about music's temporal features and its affective power on another path.

Hermeneutical considerations

The course on which I now intend to embark marks a further turning point in my reflections on music's temporal character and its affective force. The preceding discussion of the challenge of accounting for music's temporal constitution is the staging ground for the hermeneutical considerations that I set out in the rest of this chapter. Taken together, my critiques of Dahlhaus's attempt to resolve the question of the feeling of motion, Hanslick's reference to the play of tonal forms, and Hegel's insight into the difference between the succession of instantaneous "now" points and sound's sublation of external, spatial coordinates outline the course of reflections that follow. These critiques support two initial conclusions: (1) Tones that succeed one another in succession elicit a feeling of motion only by reason of the schematizing operation from which music's affective power also springs; and (2) divested of its metaphysical connotations, the spirit of a work reveals itself through the manner in which a passage or a work takes shape. This hermeneutical understanding more than just reverses the romantic conception that the spirit of a work lies somewhere behind the work, as though the work embodies the composer's intentions. Rather, identifying the spirit of a work with the work's exemplification of feelings and moods underscores how the work's expression is also the wellspring of its worlding power.

The impression created by sequences of tones or sounds is clearly attributable to the character of a pattern, figure, or line drawn from the successive combinations of sounding elements. Music's meaning and significance accordingly depend on how listeners (or test subjects in the case of cognitivist experiments) parse tones and sound elements. This widely accepted commonplace, however, eclipses the truly remarkable ability that we exercise when we hear a melody or phrase rather than a mere succession of tones, harmonies, and timbral colors. Don Ihde remarks that the "*temporal* span or duration of sounding that is experienced in listening"[41] figures prominently in forming the contours of any resulting gestalt. Yet the resulting musical gestalt is only built up, as it were, through the act of drawing a patterned figure from the melody's or phrase's individual components, as I have already indicated. Conversely, Nicholas Cook insists that listeners in Western or Westernized musical cultures normally "'see' music as moving in space."[42] However, by attributing the continuity of a musical passage to the "coherence of its movement through time,"[43] he presupposes the temporal operation on which the apprehension of a musical gestalt is based.

Listening is accordingly the act in which the auditor attends to the ways in which figures, gestures, and rhythmic patterns take shape. By insisting that the politics of listening turns on the struggle over the legitimate regime imposed by a work and by ideas about it, Peter Szendy highlights the tension between listeners' expectations and the demands made by the work.[44] The history of listening that he constructs foregrounds how concepts of the work, the composer's authority, and authenticity inform listeners' engagement with music. Listening habits thus not only anchor listeners' experiences in mores and practices governed by social values and norms, but they can also enforce rote responses, as Adorno attempted to demonstrate with his critical typology of modes of musical conduct.[45] Sander van Maas argues that, whereas Adorno's typology organizes the relation between the object of listening (the *what*) and the listening subject (the *who*) based on the distinction between them, contemporary theories of listening turn listening back on itself in order to interrogate listening practices.[46] This theoretical *(re-)tour* unsettles habituated conventions and norms. At the same time, we might wonder whether folding the act of listening back on itself vis-à-vis the listening subject's relation to audible phenomena occludes the communicability of the experience afforded by music in placing this experience in suspense.

How, we could therefore ask again, do tones, sonorities, acousmatic events, and the like constitute the object of listening if not by way of the manner in which the sounding event to which we attend takes shape as it unfolds? Listening strategies clearly play a role in listeners' experiences in this regard. Yet, apart from the demands made by the work, the experience occasioned by a work is rendered mute. The communicability of this experience depends in large part on the listener's capacity to hear tones, harmonic sonorities, percussive effects, and timbral combinations that sound in succession as giving a figure and a body to music's affective tonalities. (The exercise of this capacity is culturally contextual and tradition-dependent.) The demands that a work places on the listener's apprehension of the work's temporal expression are therefore vital to the manner in which the work speaks.

Hearing's primacy over seeing, Gadamer reminds us, rests on the dialectic that joins the receptivity of hearing with the experience of being addressed, to which we open ourselves in listening (and which in philosophies of testimony is comparable to the feeling of being enjoined in response to the Other's anteriority, exteriority, and height).[47] Martin Heidegger, moreover, remarks that "it requires a very artificial and complicated frame of mind to 'hear' a 'pure noise.'"[48] In short, "listening to..." and "hearing as..." are reciprocally related. Neither a musical work, a figure, a gesture, nor their constituent components subsist as permanent objects. On the contrary, tones, themes, and entire works vanish with their passing, as Hegel's reflections on music attest. Musical works therefore not only require their "renewed vivification"[49]; the work's performance also occasions its expression of the feelings and moods that it possesses. The question, then, is not only, "How do 'successive particles of time assemble themselves ... [in such a way as to] construct a perceivable whole'?"[50] but also "To what should we attribute the work's expressive force in the light of its fleeting temporal existence?"

The Eleatic paradox is instructive in this regard. This paradox arises from a thought experiment in which an object moving in space—Zeno's flying arrow—is represented in terms of a series of points along a line. Reduced to this series of instantaneous "now" points, the arrow's trajectory coincides with a series of isolated instances so that the arrow no longer flies. Hence the paradox. Alfred Schutz draws on this paradox to emphasize that music is not a succession of static incidents but an indivisible unit in the stream of our consciousness. He explains that like the "inner time of our stream of consciousness, [which] is entirely free from elements of space, ... we experience movement [in music] as an unbroken ongoing occurrence."[51] Consequently, he sets the flux of music's temporally indivisible unity against the physicist's or the philosopher's conception of cosmic time.

For Schutz, the Eleatic paradox is proof that music is an event that exhibits all the criteria of an ongoing motion that can be broken up only at the cost of stepping out of the flow of its temporal development. Measuring motion necessarily makes reference to spatial coordinates in order to set the trajectory traced out by music's temporal movement against the fixed spans traversed by its ongoing flow (as I indicated in my previous discussion of Dahlhaus's recourse to a *temps durée* and a *temps espace*). But then the Eleatic paradox once again confounds the experience of music's inner temporal flux, since this flow would be arrested through being broken up into an infinite series of discrete instances. Schutz consequently refuses to identify music's inner time with time's spatial extension, as is the case when inner time is projected spatially so that human action takes place in a public dimension. For him, the time that we experience in listening to music is therefore incommensurable both with ordinary time (as measured by the movement of the clock) and the time of the world.

The entire difficulty of accounting for music's temporal movement springs from the incommensurability of these two orders of time: cosmic time and the time internal to consciousness. Time that can be measured in relation to the motions of the heavenly bodies is external to the experience of music's internal movement. Conversely, time that is experienced as an ongoing stream in its indivisible unity is internal to our consciousness. (In the next chapter, I will review the aporias that arise from the attempt to constitute this internal time-consciousness independently of any reference to the ordinary concept of time.) For Schutz, the "dimension of time in which the work of music exists is the inner time of our stream of consciousness."[52] The *temps durée* (in Bergson's terminology) is therefore built up polythetically as each successive impulse contributes to structuring music's ongoing temporal flux. For Schutz, music's polythetic constitution is the basis of music's own inner temporal unity. Consequently, this inner unity is a function of the configurating operation that draws elements together temporally. By putting into play the faculties of retention (memory) and protention (anticipation), the polythetic process is one through which a work's successive figuration unfolds. Schutz can thus liken points of tension and relaxation to cresting waves that follow one another so that each impulse "is experienced as a unit, [that is, as] a movement in inner time."[53]

I readily admit that my remarks on the way that a work is built up overstep Schutz's claim that the object of musical perception owes its ideality to a polythetic principle. Schutz is right to insist that the "listener experiences the theme [of a work] as an indivisible unit."[54] Yet a theme or a melody owes its coherence to the way that the listener draws a figure from a series of individual tones. The melody confers a sense of temporal unity of this series of tones. The polythetic principle that Schutz identifies as vital to the listener's perception of music's temporal flux therefore presupposes the operation that draws a succession of tones together in their melodic configuration. Schutz emphasizes that we have a feeling that a certain configuration is responsible for structuring music's ongoing flux.[55] The ideal object that he identifies with music's polythetic constitution not only rests on the fact that a theme starts, develops, and recurs, but this purportedly ideal object also relies on the listener's capacity for following the theme as it unfolds. Like Gabriel Marcel, who insists that a melodic line's aesthetic attributes are features of a non-spatial figure that "transcend … the purely temporal mode according to which it appears,"[56] Schutz's contention that music's meaning cannot be recuperated conceptually reinforces the idea that music's expressions are irreducible to "fixed" atemporal schemata. Consequently, expression in music springs from the way that temporally ordered sounds and silences give voice to the affective properties of a work, an extemporized performance, or even an "empty" duration.

By identifying music's power of expression with the process through which successive heterogeneous elements give rise to temporally unfolding figures, I want to emphasize how music's temporal character differs from, and is irreducible to, a series of sounding events plotted out along a spatial axis. The percussive noises in Edgar Varèse's *Ionisation*, the serially controlled ordering of pitches in Pierre Boulez's *Structures 1a*, the timbrally modulated qualities of a single tone in Elliot Carter's *Etude No. 7* (from *Eight Etudes and a Fantasy for Woodwind Quartet*) or the ambient sounds that populate John Cage's "silent" work, *4'33"* (the title of which is taken from David Tudor's 1952 première performance, in which the durations of the three "movements" were 33", 2'40" and 1'20" respectively)—all of these only figure as meaningful or significant elements—intended or not—insofar as noises, pitches, shifting timbral qualities, or happenstance sounds shape the ethea of these different works.[57] Just as music's temporal features spring from the way that successively sounding elements cohere, the figures—motives, themes, gestures, movements, and even entire works—that crystallize as they unfold give a body to their distinctive attributes. In each case, the configuration of various elements confers a specifically temporal character on sequentially ordered successions of sounds (even when this ordered succession is the result of chance occurrences). A theme's, a gesture's, or a work's temporal configuration thus evinces the productive character of the operation that transforms the rectilinear ordering of individual tones into a unique, sounding passage. Apart from this operation, motives, themes, rhythmic patterns—in short, music—would not rise to the level of patterned expressions or "humanly organized sound."[58] On the contrary, music would be devoid of any affective value, since like words that could not be combined to form sentences, individual sounds would scatter in the void of meaningless reverberations.

In attributing music's expressive force to its temporal configuration, I want to make clear how the process through which figures and gestures take shape is also the engine of music's worlding power. Music's power to refashion our manner of inhering in the world, I will argue shortly, has its root in music's mimetic character. Music's revaluation of time in response to the deficiency that I indicated earlier bears the stamp of the difference between time and eternity is proof, so to speak, of music's mimetic refiguration of time's ultimate unrepresentability. Expression, which I stressed is indistinguishable from the process through which figures and gestures acquire their distinctive character, is therefore also the spring of a work's power to refashion our affinities with the world. Expression thus not only signals the entry of aesthetics into the field of phenomenological hermeneutics; since expression is emphatically temporal (as is evident in the fact that the work's expression consists in its successive configuration), a work's expression also attests to its irreducibility to the ordinary concept of time.

Musical analysis takes music's temporal configuration for granted in identifying integrative systems or networks of relations that are said to constitute the work's formal coherence. When in Chapter 7 I relate a work's exemplification of feelings and moods to the work's claim to truth, I will show how the matrix that constitutes the schema of a work's system of organization is corollary with the configurating operation to which I have credited music's expressive force. This matrix constitutes the density of the aural field in which a work's affective tonalities are felt and heard. By schematizing the synthetic operation that transforms a succession of tones and sonorities into melodic and thematic figures, this matrix is thus the ground of the formal schemata on which musical analysis depends. Analysis, which Medieval Latin derived from the Greek *analuein*, to unloose, necessarily detemporalizes the systems of relations that it identifies. Analytic abstractions thus provide second-order accounts of a work's structural features and properties based on a temporally dynamic network of interrelationships. The ideal of organic unity, which was widely applied as a metaphysical precept in the nineteenth century and then again as an analytic principle in the twentieth century as the mark of a work's value and greatness, traded on the formal integrity attributed to the way that a work's constitutive parts fit naturally or logically together. However as I have indicated, the dynamic relation between parts and whole, which was said to emulate a living organism's processes of growth and development, is itself a function of the work's temporal configuration.[59] Adorno, who regarded the musical work as a *force-field* (*Kraftfeld*) organized around a challenge or problem, rightly attributes the work's logic to the set of dynamic relationships that contained the problem animating the work.[60] His virulent critique of philosophies of identity unleashed a wholesale renouncement of every ontology. This critique notwithstanding, the logic of a work for him retains its connection to the dynamic complexes constituted by the work as a temporal force field.

Spatializing representations of music's temporal schemata clearly enable the analyst to break a work apart in order to identify salient structural features. Viewing a work as a system of fixed synchronic relations, which can be scrutinized independently of their specifically temporal attributes, is methodologically justified as a means of setting the work's unique features against the backdrop of nomic generalities. The logical space mapped out by the set of conventions that govern the structure of harmonic progressions in tonal compositions, for example, assumes its quasi-universal character by reason of an abstract system drawn from actual tonal examples. However, the logical space of possible tonal relationships exists independently of any particular work only as a theoretical construct. Consequently, the notion that this logical space comprises the "framework within which, and through which, the actual sequence of events is shaped"[61] reverses the order of priority of theory and analysis vis-à-vis music's temporal constitution. Since the system of tonal relations can be represented abstractly, it could be said to exist independently of any particular instantiation in a composition. The strength of Heinrich Schenker's theory, which ostensibly lies with the way that it "defines this space in terms of a conventionalized, unchanging vertical background that is rendered linear through a set of equally fixed operations, ultimately producing the richness and relational complexity of a particular musical composition,"[62] is rooted in this power of abstraction. Robert Morgan therefore concludes that apart from the concept of a tonal space, a system of structured interrelations in music would be unthinkable. Analysis, however, stands in the same relation to composition (literally, arranging the parts by assembling them together) as theory does to the practices on which music theory is based. Consequently, spatial projections derived from the schematization of music's unfolding structures lay out the work's temporal matrix along a rectilinear grid.

Music's affective properties, however, come to life only in the experience occasioned by a work's performance. Correlatively, feelings and moods take shape in accordance with the way that a theme, a passage, or a work unfolds. Visceral imitations of musical gestures in bodily reactions are visible signs of a listener's internalization of the work's affective features, which the work's expression renders communicable.[63] The affective tonalities exemplified by a work come to expression only through their temporal figuration. Hence, these affective tonalities are properties of the temporal matrices drawn from the way that tones, harmonies, and noises heard in succession cohere.

The opening passage of Bach's "Prelude VII" from *The Well-Tempered Clavier* (BWV 852; Example 1.1) evinces this connection between a passage's affective qualities and its temporal configuration. From the standpoint of the Baroque doctrine of affects, the opening figure of Bach's "Prelude" (m1) might be regarded as an instance of *anabasis*.[64] As such, this initial figure could be identified as representing the feeling of a rising expectation accentuated by the ascending e♭′-d♭″ leap. The affective quality of this figure, however, is a feature of the tones that sound in succession (g′-a♭′-b♭′-c″-b♭′-a′-g′-f′-e♭′-d♭″). Hence, this affective quality is a property of this figure's temporal unity. The ascending e♭′-d♭″ leap inflects this opening passage harmonically by transforming the tonic triad into the subdominant's dominant seventh (V^7/IV) over a sustained e♭ tonic pedal. Together, this figure's recurrences and the chain of suspensions that tie them together embody affective attributes that theories of imitation (such as the doctrine of affects) identify with music's representational value. The succession of tones therefore constitutes only the dimension of this passage's temporal extension; the passage's expression owes its force to its temporal configuration as a whole.

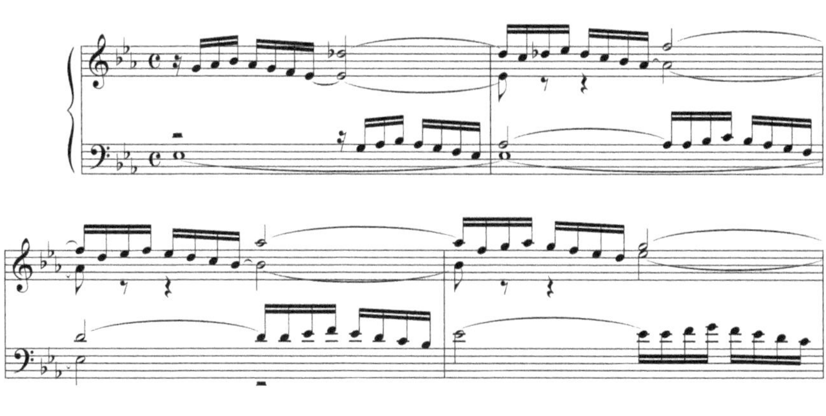

Example 1.1 J. S. Bach, "Prelude VII" from *The Well-Tempered Clavier* Vol. 1, mm. 1–4.

Attributing music's expression to the dynamic relation between successively sounding tones, harmonic sonorities, timbral combinations, and the like and a theme's, a passage's, or a work's temporal configuration marks a major difference between the philosophically hermeneutical approach that I am taking and interpretive approaches that in one way or another privilege representations in music of so-called extramusical meanings. Hermann Kretzschmar, who famously regarded music as a "born auxiliary art,"[65] deployed musical hermeneutics as an exegetical method for aiding the listener's comprehension. Drawing on the aesthetic doctrine of affects, he supplemented music's supposed deficiency with respect to language's representational power with narrativizing explanations of the work's expressive content. Hence, he adhered to the epistemic shift that privileged an order of knowledge where truth was defined in terms of the adequacy of the representation and that allied music's figurative expressions with the Baroque doctrine of affects. With the advent of this new order of knowledge, musical figures represented feelings, affects, and passions in accordance with the theory that these figures imitated the rhetorical use of language.[66] Representational theories and theories that subscribe to the idea that music is a fundamentally nonrepresentational art are paradoxically beholden to the concept of language espoused by representative thought "when it treats language as *Ausdruck*, 'expression'—that is, as the exteriorization of the interior, and hence as the domination of the outside by the inside."[67] By drawing out the connection between music's temporal configuration and its expression of feelings and moods to which individual works give voice, my own approach to the question of music's expressive power and ontological vehemence breaks with metaphysical and epistemological tacks. Accordingly, my thesis that music revalues time in response to time's deficiency vis-à-vis its other brings to the fore music's mimetic function.

The mimetic function

The break with metaphysical, epistemological, and sociological assumptions regarding music's representational value distinguishes my hermeneutical perspective from romantic sensibilities, postmodern deconstructions, and cognitivist accounts of music's emotive force.[68] In the preceding discussion, I have gone to some lengths to explain how the transformation of successively sounding elements into temporally expressive figures gives rise to music's affective properties. In the final chapter, I will explore in greater detail the connection between the schematizing operation that draws together the successively sounding elements that produce these figures and the communicability of the experience occasioned by a work. For now, I only want to emphasize how a work's temporal configuration is also the condition of its worlding power.

The work's retreat from the practical order of our everyday life, it turns out, is the condition for music's subversion of congealed practices and habits of feeling. The difference between the way that works place the real into suspense and the metaphysical value vested in music through its institution as an aesthetically autonomous entity is critical in this respect. As I will explain further in Chapter 7, the distance that works take from the real can no more be credited to the way that art acquired its own autonomous standpoint in the nineteenth century than can music's and art's productive potential be derived from the aesthetics' ideologically deleterious function. Like literature, theater, and visual and performance art, music would be completely innocuous apart from its power to shatter existing habits and practices through laying claim to new and different ways of feeling, thinking, and acting. From this hermeneutical vantage point, the work's distance from the real is a function of the work's imaginative explorations. This distance from the real is therefore the mark of the rupture between the world projected by a work and the practical field of our everyday experiences. This world, which unfolds in accordance with the work's expression of it, gives voice to feelings and moods that are unique to the work. Through closing itself off from the practical order of reality, music acquires its affective force through exemplifying modalities of feeling that did not exist prior to the work's evocation of them.

The transposition of everyday feelings and emotions, which music brings about by giving voice to the feelings and moods that it expresses, marks out the domain of music's mimetic function. Like poetic language, which inverts the way that ordinary language is directed outward in order to structure the mood that is interior to a poem, the feelings and sentiments expressed by music are coextensive with the work's temporal configuration.[69] As is the case with poetry, narrative, and indeed all of the arts, feeling's elevation to fiction in music is the condition of feeling's mimetic deployment. By constructing a "world of singular essences in the realm of feeling,"[70] music raises sentiments and moods that bear on the ways that we are attuned to ourselves, others, and the world to the level of fiction in uniquely singular ways. It is as though by abolishing art's and language's representational function, music gives a figure and a body to states of mind that are manifested in, and that resonate with, the affective qualities of music's temporal schemata.[71] Music's power to move us and impassion our desires attests to the force with which music touches the heart of our incarnate existence as body and flesh.[72] In answer to the absence of aseity born from the difference between time and the alterity of an order that surpasses time, music reverberates with feelings that revalue time through renewing our inherence in the world.

In this regard, music differs from the art of narrative. Narrative draws upon the resources of language and its different tenses—preterite, present perfect, and future indicative, for example; music evokes feelings and moods in accordance with the work's temporal expression. The "thought" or "idea" (*dianoia*) expressed by the story as a whole thus has its counterpart in the affective properties embodied—or better, iconized—by the musical work's temporal matrix. (Conversely, the musicality of a text or poem is equally attributable to the way that the text or poem displays its feelings or moods by schematizing them.) By drawing a distinction between narrative's diegetic value and the mimetic character that she identifies with music's ephemeral character, Carolyn Abbate emphasizes the difference between the narrative art's practical resources and music's expressive power. By insisting that music's mimetic impulse "traps the listener in present experience and the beat of passing time,"[73] she also provides an indication of how music's temporal character abolishes the difference between narrative time (*Erzähltezeit*) and narrated time (*Erzählte Zeit*). Abolishing the difference between the "fact of 'narrating' and the thing 'narrated'"[74] in a story's presentification, however, does not mean that we should embrace the supposed immediacy of music's temporal flux. On the contrary, as I have argued, music's temporal configuration is the requisite condition of the expressive vehemence and ontological force that I have identified with music's worlding power.

When in the Introduction I highlighted the difference between music's mimetic significance and that of the art of narrative, I anticipated this conclusion. By breaking with the time with which we reckon in our daily lives, musical works, like narrative fictions, lift themselves "from the opaque depths of living, acting, and suffering."[75] This break with the time of our everyday lives is the law governing the entry into the kingdom of fiction. Yet the closer the art of narrative draws to the inscrutability of time, the nearer this art approaches its limits. If lyric poetry gives voice to the grief fed by the miseries of life, music resounds with the pathos born from the deficiency that strikes at the heart of our experiences of time. This deficiency is the sign of the difference between time and the alterity of an order to which the art of narrative points through exceeding itself to the point of its exhaustion. In exploring the borderlines between time and eternity, music sounds out this difference in response to the enigma of time's inscrutability. Hence, music's sublimation of the deficiency that strikes at the heart of our experiences of time opens new regions of feeling. As such, music's mimetic refiguration of time's ultimate unrepresentability renews our manner of inhering in the world through refashioning our affective affinities with it.

My reflections on music's expressions of its temporal and affective qualities are guided by the idea that music opens the world to us and us to the world in response to questions, perplexities, and crises regarding the meaning of time. The strategy that I pursue in the chapters devoted to exploring the significance of different works draws support from the analysis in the next chapter of the ontological deficiency brought to light by a meditation on the being and the non-being of time. Ultimately, I hope to show how music revalues time through staking out the boundaries between time and eternity. It remains to be seen how different works reply to the ontological deficiency that places its negative stamp on time. At the least, music's evocation of an experience that approaches its own limit the closer it draws to the enigma of time's other stands as witness to the enduring inscrutability of a phenomenon that everywhere and always envelops us.

Notes

1 Carl Dahlhaus, *Esthetics of Music*, trans. William Austin (New York: Cambridge University Press, 1982), 75.
2 Dahlhaus, *Esthetics*, 75.
3 Ibid., 78.
4 Ibid., 74–75.
5 Ibid., 77.
6 See Paul Ricoeur, "What is a Text? Explanation and Understanding," in *Hermeneutics and the Human Sciences: Essays on Language, Action and Interpretation*, trans. John B. Thompson (Cambridge: Cambridge University Press, 1981). I am aware that the experiences occasioned by these other arts also have their temporal characteristics. Music is exemplary in this regard.
7 For remarks on music as a *poiesis a se*, see Ernst Bloch, *The Principle of Hope*, vol. 3, trans. Neville Plaice et al. (Oxford: Basil Blackwell, 1986), 1069.
8 Dahlhaus, *Esthetics*, 81; see Roger Scruton, *The Aesthetics of Music* (Oxford: Oxford University Press, 1997). Scruton remarks that "[s]patial metaphors permeate our experiences of music, and the organization which produces music out of sounds prompts us … to think of sound in spatial terms" (15). At the same time, he stresses that "[w]hen we hear music, we do not hear sound only; we hear something *in* the sound, something which moves with a force of its own" (19–20).
9 Charles Keil, "Participatory Discrepancies and the Power of Music," *Cultural Anthropology* 2, no. 3 (1987); Charles Keil, "The Theory of Participatory Discrepancies: A Progress Report," *Ethnomusicology* 39, no. 1 (1995).
10 Eduard Hanslick, *On the Musically Beautiful: A Contribution toward the Revision of the Aesthetics of Music*, trans. and ed. Geoffrey Payzant (Indianapolis, IN: Hackett Publishing Co., 1986), 78. Tellingly, architecture offers a lofty analogy for the musically beautiful only by relinquishing the "attribute of motion" (29), thereby stripping away the expressly temporal character of musical ideas that take shape through the play of these tonal forms.
11 Hanslick, *Musically Beautiful*, 8–9. Hanslick's adoption of a scientific standpoint thus combats the gnostic character of absolute music's purely expressive qualities. See Mark Evan Bonds, *Absolute Music: The History of an Idea* (Oxford: Oxford University Press, 2014). Bonds points out that despite Hanslick's insistence that "music exists entirely for its own sake, Hanslick acknowledges that it can … provide an 'intimation' of the highest idea of all, the Absolute" (145). As I will argue, Hanslick finally could not extricate his aesthetics from romantic influences.
12 Hanslick, *Musically Beautiful*, 81.
13 Hanslick, *Musically Beautiful*, 81. Hanslick emphasizes that if the "artist is inscrutable, the artwork is decidedly not" (33).
14 Hanslick, *Musically Beautiful*, 29; see Bonds, *Absolute Music*. In his exegesis of Hanslick's phrase *tönend bewegte Formen*, Bonds remarks that "Hanslick's 'forms' (*Formen*) are set in motion (*bewegte*) by 'sounding tones' (*toned*)" (147). Since the "sound of humanly produced forms … creates forms that move through time" (147), *Formen*, for Bonds, are more appropriately understood as manifestations or configurations.
15 Hanslick, *Musically Beautiful*, 30. For a discussion of Hanslick's subsequent rejection of his positivist program, see Kevin C. Karnes, *Music, Criticism, and*

the Challenges of History: Shaping Modern Musical Thought in Late Nineteenth-Century Vienna (New York: Oxford University Press, 2008).

16 Hanslick, *Musically Beautiful*, 31. That Hanslick is unable to account for the difference between these laws and changing tastes calls into question these laws' ostensible immutability.

17 Hanslick, *Musically Beautiful*, 82. See Hans-Georg Gadamer, *Truth and Method*, 2nd rev. ed., trans. Joel Weinsheimer and Donald G. Marshall (New York: Crossroad Publishing, 1989), 101 ff.; Max Paddison, "Mimesis and the Aesthetics of Musical Expression," *Music Analysis* 29, no. 1–3 (2010): 132 ff. Paddison suggests that by including the element of play, Hanslick is attempting to deepen the concept of mimesis that Paddison identifies with inner impulses associated with the movement of music's sensuous materials.

18 Gadamer, *Truth and Method*. Gadamer explains that "play expresses itself in the unity of its course" (113). He can therefore claim that:

> Play is structure … [since] despite its dependence on being played it [structure] is a meaningful whole which can be repeatedly presented as such and the significance of which can be understood. But structure is also play, because—despite this theoretical unity—it achieves its full being only each time it is played.
>
> (117)

19 Hanslick, *Musically Beautiful*, 31. By crediting music's ideal content to the imagination's spontaneous grasp of tonal connections, Hanslick tacitly ascribes the logic of a work to the way in which the play of tonal forms crystallizes. This logic is accordingly a function of the operation that draws a configuration from a succession of tones.

20 See Lydia Goehr, *The Imaginary Museum of Musical Works: An Essay in the Philosophy of Music* (Oxford: Oxford University Press, 1992); Nicholas Cook, "Introduction: Refocusing Theory," *MTO: A Journal of the Society for Music Theory* 18, no. 1 (2012), accessed November 10, 2016, http://mtosmt.org/issues/mto.12.18.1/mto.12.18.1.cook.html; Daniel Leech-Wilkinson, "Compositions, Scores, Performances, Meanings," *MTO: A Journal of the Society for Music Theory* 18, no. 1 (2012), accessed November 10, 2016, www.mtosmt.org/issues/mto.12.18.1/mto.12.18.1.leech-wilkinson.php.

21 Hanslick, *Musically Beautiful*, 33. Hanslick adds that a musical passage has "its effect upon the listener in this way and no other" (33). See Eldritch Priest, "Felt as Thought (or, Musical Abstraction and the Semblance of Affect)," in *Sound, Music, Affect: Theorizing Sonic Experience*, ed. Marie Thompson and Ian Biddle (New York: Bloomsbury, 2013). In his reading of Susanne Langer's theory, Priest argues that "[u]nlike language, whose arbitrarily assigned signs make its symbolic function explicit, music is typically not regarded in a way that distinguishes its meaning from its figures" (57).

22 Hanslick, *Musically Beautiful*, 32.

23 Ibid., 82.

24 Hanslick, *Musically Beautiful*, 31; see Daniel K. L. Chua, *Absolute Music and the Construction of Meaning* (Cambridge: Cambridge University Press, 1999). Chua identifies the excised passage as referring to music as the "sounding image of the universe" (quoted in Chua, 230).

25 Carl Dahlhaus, *The Idea of Absolute Music*, trans. Roger Lustig (Chicago, IL: University of Chicago Press, 1989), 152–153. Dahlhaus points out that

> Hanslick adopted the difference between idea and appearance in order to be able to define the beautiful in music…; unlike Hegel, however, he did not determine the sounding phenomenon to be appearance and "thoughts and

> feelings" to be the idea or (as Hegel said) content, but rather sought the idea or content in the specifically musical aspect.
>
> (110)

See Bonds, *Absolute Music*, 143 ff.

26 Karnes, *Music, Criticism and the Challenge of History*, 53. According to Karnes, this motivation propelled Hanslick to write his *History of Concert Life*. The paradox that an objective science could never reach or penetrate subjective judgments regarding a work's meaning and significance presaged Hanslick's epiphany in the 1860s, which led him to reject his positivist approach.

27 Georg Wilhelm Friedrich Hegel, *Hegel's Aesthetics: Lectures on Fine Art*, vol. 2, trans. Thomas Malcolm Knox (Oxford: Oxford University Press, 1975), 88.

28 Hegel, *Aesthetics*, 909.

29 Ibid., 26.

30 Ibid., 891.

31 Ibid.

32 Ibid., 890.

33 Ibid., 902.

34 Ibid.

35 Hegel, *Aesthetics*, 907. For Hegel, the "actual self belongs to time" (907) with which it coincides. Hence, the "self is in time, and time is the being of the subject himself" (908).

36 Hegel, *Aesthetics*, 890:

> [T]he negativity into which the vibrating material enters here is on one side the cancelling of the spatial situation, a cancellation cancelled again by the reaction of the body, therefore the expression of this double negation, i.e. sound, is an externality which in its coming-to-be is annihilated again by its very existence, and it vanishes of itself.

37 Hegel, *Aesthetics,* 906 (original emphasis). Music is "concerned only with the completely indeterminate movement of the inner spirit and with sounds as if they were feeling without thought" (28). Hence, the subjective inner life that music takes as its subject matter is presented as free from all external constraints. For Hegel, beauty is the sensuous presentation in art of Spirit [*Darstellung*] in which the Idea is sensibly manifested, thereby proceeding to a deeper and more inward expression of the inner life of a self-conscious subjectivity. In art, the free self-consciousness of Spirit comes to know itself as other than nature. Accordingly, music withdraws from all external reality to the inward recesses of subjective feeling. By canceling first the spatial plane of the painted canvas, and second its own external materiality, music supersedes painting as a higher Romantic art. Hearing for Hegel is thus a more ideal sense than sight because in listening to music, one apprehends "the first and more ideal breath of the soul" (890). In music, the double negation of all external reality in the temporal movement of sound becomes adequate to the expression of the inwardly self-conscious life of spirit in its subjective freedom.

38 Hegel, *Aesthetics*, 1. In the *Phenomenology of Spirit*, Hegel announces that the "True is the whole. But [then] the whole is nothing other than the essence consummating itself through its development" (Georg Wilhelm Friedrich Hegel, *Hegel's Phenomenology of Spirit*, trans. A. V. Miller [Oxford: Oxford University Press, 1977], 11). According to Charles Taylor, Spirit's embodiment in history "is not only the fulfillment of his rationally necessary plan, the Idea, but also the paradigm expression of what that purpose is" (Charles Taylor, *Hegel* [Cambridge: Cambridge University Press, 1975], 469); see Hans-George Gadamer, *The Relevance of the Beautiful and Other Essays*, ed. Robert Bernasconi, trans. Nicholas

Walker (Cambridge: Cambridge University Press, 1986). Gadamer remarks that "in Hegel's eyes, this task and this claim, which philosophy has always made, are only fulfilled when philosophy comprehends and gathers up into itself the totality of truth as it has been unfolded in its historical development" (5).

39 Hegel, *Aesthetics*, 906 (original emphasis): "[T]he result is that music with its movements penetrates the arcanum of all the movements of the soul."

40 Hegel, *Aesthetics*, 28 (my emphasis).

41 Don Ihde, *Listening and Voice: Phenomenologies of Sound*, 2nd ed. (Albany, NY: State University of New York Press, 2007), 89 (original emphasis).

42 Nicholas Cook, *Music, Imagination and Culture* (Oxford: Oxford University Press, 1990), 218. For Cook, the ideal of a work's organic integrity thus "forms the cornerstone of the *construction* of a modernist regime of listening" (127). According to Cook, differences between perceptual experiences and metaphorical representations of music based on theoretical precepts constitute the pattern of divergences that constitute a musical culture's identifying features. By drawing out the prejudices and presuppositions that dominate Western (or Westernized) musical cultures, he accordingly stresses the significance of these prejudices and presuppositions in constituting conglomerations of sounds as intentional objects.

43 Cook, *Music, Imagination and Culture*, 28; see Nicholas Cook, *Beyond the Score: Music as Performance* (New York: Oxford University Press, 2013). By shifting the focus of discourses on music's meaning away from the score's authoritative value towards music's performative character, Cook emphasizes that the experience of live or recorded music is fundamental to any inquiry into music's power and force. "Plato's curse," to which Cook attributes the twinned ideals of the musical work's conception as an abstract entity and its status as a fixed notated entity, reinforces the difference between textual interpretations and analyses of music's performative character. Cook's antipathy toward an ontological approach that qualifies each performance in terms of its adherence to its original meaning thus opens the door to a phenomenologically hermeneutical understanding of music's temporal configuration.

44 Peter Szendy, *Listen: A History of Our Ears*, trans. Charlotte Mandell (New York: Fordham University Press, 2008); see Peter Szendy, "The Auditory Re-Turn (The Point of Listening)," in *Thresholds of Listening: Sound, Technics, Space*, ed. Sander van Maas (New York: Fordham University Press, 2015).

45 Theodor W. Adorno, "Types of Musical Conduct," Chapter 1 of *Introduction to the Sociology of Music*, trans. E. B. Ashton (New York: Continuum Publishing Company, 1976).

46 Sander van Maas, "Introduction," in *Thresholds of Listening: Sound, Technics, Space*, ed. Sander van Maas (New York: Fordham University Press, 2015), 4. Theories of listening occupy a prominent place in contemporary discourses. See Cook, *Music, Imagination and Culture*; David Clarke and Eric Clarke, eds., *Music and Consciousness: Philosophical, Psychological, and Cultural Perspectives* (Oxford: Oxford University Press, 2011); Anthony Gritten and Elaine King, *Music and Gesture* (Burlington, VT: Ashgate, 2006); Keith Chapin and Andrew H. Clark, eds., *Speaking of Music: Addressing the Sonorous* (New York: Fordham University Press, 2013); Veit Erlmann, ed., *Hearing Cultures: Essays on Sound, Listening, and Modernity*, (Oxford: Berg Publishers, 2004).

47 Gadamer, *Truth and Method*, 462. Throughout *Music, Time, and Its Other*, I distinguish between claims stemming from musical works and injunctions issuing from exemplary moral and political acts. See especially Chapters 3 and 7.

48 Martin Heidegger, *Being and Time*, trans. John Macquarrie and Edward Robinson (New York: Harper and Row, 1962), 207. For Heidegger, acoustic

perception is based on hearing. Accordingly, "[l]istening to … is Dasein's existential way of Being-open as Being-with for Others…. Dasein hears because it understands" (206).

49 Hegel, *Aesthetics* vol. 2, 909; see Gadamer, *Truth and Method*. Gadamer emphasizes that it is only in performance "that we encounter the work" (16). Hence, music "must resound" (ibid.). Hanslick similarly maintains that music is a "resounding actuality" (Hanslick, *Musically Beautiful*, 49).

50 Hanslick, *Musically Beautiful*, 69.

51 Alfred Schutz, "Fragments on the Phenomenology of Music," ed. Fred Kersten, *Music and Man* 2, no. 1–2 (1976): 37; see David Clarke, "Music, Phenomenology, Time Consciousness: Meditations after Husserl," in *Music and Consciousness: Philosophical, Psychological, and Cultural Perspectives*, ed. David Clarke and Eric Clarke (Oxford: Oxford University Press, 2011). Clarke argues that music mirrors the way that consciousness apprehends itself in the unity of its flux by molding, modelling, and making "audible the flow of our inner, subjective life" (1). Eric Clarke similarly identifies musical consciousness with the dynamically changing awareness of a stream of auditory perceptions (Eric Clarke, "Music Perception and Musical Consciousness," in *Music and Consciousness: Philosophical, Psychological, and Cultural Perspectives*, ed. David Clarke and Eric Clarke [Oxford: Oxford University Press, 2011]). See also Lawrence M. Zbikowski, "Music, Language, and Kinds of Consciousness," in *Music and Consciousness: Philosophical, Psychological, and Cultural Perspectives*, ed. David Clarke and Eric Clarke (Oxford: Oxford University Press, 2011). Zbikowski identifies musical consciousness with the "cascade of mental images" (179) prompted by a performance in response to the musical events that succeed one another in time. Consciousness of the way that these cascading images steadily unfold over time rests on the capacity to follow the flow of successive events. Consequently, the intentional object of this consciousness is constituted by the operation that draws successive events together as a temporal whole. See also Lawrence M. Zbikowski, "Music, Emotion, Analysis," *Music Analysis* 29, no. 1–3 (2010). For a discussion of the aporias engendered by the phenomenology of time, see Chapter 2.

52 Schutz, "Fragments," 31. According to Schutz, Bergson's philosophy

> has taught us, in full clarity, the double aspect of movement: On the one hand, movement means the ongoing motion, the perceptual change of place of an object, an event which happens in time and which can only be grasped as a unity by our sense of inner time—the *durée*. On the other hand, for the movement as an ongoing occurrence our mind substitutes the spatial trace traversed by the moving object. This is a conceptual scheme, incompatible with the pure *durée*. It requires the exteriorization of *durée* into spatial time, into the time of our clocks, of our life with others—the time in which the earth turns around its axis and completes its course around the sun.
>
> (30)

53 Schutz, "Fragments," 65–66; see Christina A. Skadra, "Alfred Schutz's Phenomenology of Music," *The Journal of Musicological Research* 3, no. 1–2 (1979). Skadra explains that for Schutz, the

> work of music exists as an audible object within the dimension of inner time…. While immersed in the musical process itself the recognition of a recurrent theme as the same or different occurs not through a passive synthesis of recognition, but through a *step by step*, polythetic re-experiencing of steps once experienced and now re-experienced.
>
> (108)

Consequently, in accordance with inner experience's characterization "as a continual flux, as the irreversible stream of consciousness.... the dimension of inner time allows only for polythetic recognition" (108).

54 Schutz, "Fragments," 60. Accordingly for Schutz, the "flux of music and the flux of the stream of our consciousness are interrelated, are simultaneous, [such that] there is a unity between them" (31). See, Skadra, "Schutz's Phenomenology of Music." Skadra notes that

> [m]ovement belongs to that sphere of experience which has been called "outer" and would thus appear to be an element foreign to the experience of music—an experience of inner time. To be sure, Schutz maintains that movement is foreign to musical experience.
>
> (105)

55 Schutz, "Fragments," 31; see Roman Ingarden, *The Work of Music and the Problem of Its Identity*, trans. Adam Czerniawski, ed. Jean G. Harrell (Berkeley, CA: University of California Press, 1986). Ingarden identifies the time structured by the arrangement of a work's parts as a "structured duration" (90). He emphasizes that the "temporal coloring of the specific moments, as well as the types of time-flow, play a large role in the constitution ... of the emotional qualities in the musical work" (90). Hence for him, the

> emotional qualities in musical works are so specifically merged with sound-constructs that this is reflected in their coloring in a way not found anywhere outside music. We may therefore conclude that they are *specifically musical*, emotional qualities possessing such force and dynamic presence that they are more "affective" than in any of the other arts.
>
> (98)

Ingarden concludes that these affective qualities are "exclusive to music.... [although], it is [sometimes] mistakenly said ... that they constitute 'extra-musical subjects'" (98). See also Carl Dahlhaus, "Fragments of Musical Hermeneutics," trans. Karen Painter, *Current Musicology* 50 (1992): 17; Dahlhaus, *Esthetics of Music*, 81 ff. In contrast to Ingarden's assertion that the musical work has only one level, Dahlhaus maintains that the logic of music's inner dynamic constitutes a middle ground between music's syntactical and semantic elements.

56 Gabriel Marcel, "Bergsonism and Music," in *Reflections on Art: A Source Book of Writings by Artists, Critics and Philosophers*, ed. Susanne K. Langer (Baltimore, MD: The John Hopkins Press, 1958), 146.

57 See Mikel Dufrenne, *The Phenomenology of Aesthetic Experience*, trans. Edward S. Casey et al. (Evanston, IL: Northwestern University Press, 1973). Dufrenne stresses that the quality of feeling expressed by a work emanates from the ensemble of elements that produce the work's "world atmosphere" (178). Karlheinz Stockhausen's account of how time passes overlooks how the temporal durations derived from serially generated "formant" structures in *Gruppen*, for example, modulate the work's temporal constitution only when these durations are sounded out by the materials that fill successive and overlapping temporal spans as played by *Gruppen*'s three orchestras. Karlheinz Stockhausen, "...how time passes...," *Die Reihe* 3 (1975); see my *Structure and Sorcery: The Aesthetics of Post-War Serial Composition and Indeterminacy* (New York: Garland, 1989), 40 ff.

58 John Blacking, *How Musical Is Man?* (Seattle, WA: University of Washington Press, 1973), 32; original in italics.

59 For a critique of the justificatory function of music theory and analysis vis-à-vis the ideology of organicism, see Joseph Kerman, "How We Got into Analysis, and

How to Get Out," *Critical Inquiry* 7, no. 2 (1980); Joseph Kerman, *Contemplating Music* (Cambridge, MA: Harvard University Press, 1985). Moreover, Kerman reproaches formalist analysis for its myopic perspective, which focuses on a work's internal relationships at the expense of any broader critical consideration of the work's contextual meaning and significance (73). See also Ruth Solie, "The Living Work: Organicism and Musical Analysis," *19th Century Music* 4, no. 2 (1980).

60 Theodor W. Adorno, "On the Problem of Musical Analysis," trans. Max Paddison, *Music Analysis* 1, no. 2 (1982). For Adorno, the task of analysis is to make the work's problem explicit in order to pinpoint the "'impossible' that every piece of music wants to make possible" (181).

61 Robert P. Morgan, "Musical Time/Musical Space," *Critical Inquiry* 6, no. 3 (1980): 529.

62 Morgan, "Musical Time/Musical Space," 532; see Karnes, *Music, Criticism and the Challenge of History*. Karnes points out that in

> recent decades, Schenker's contribution to music analysis has been widely regarded as laying the foundation for an array of radial structuralist experiments.... But this image of Schenker must be complicated, to say the least, by our recognition of the fact that his first attempts at analytical work revealed a profound distrust of the influence of positivist scholarship upon the crucial discourse of music.... The critic, Schenker held, must embrace subjective impressions, indulge the hermeneutic impulse, and even probe the depths of the creative mind in his attempts to elucidate the effectiveness and worth of the artworks he considers.
>
> (107–108)

Karnes therefore suggests that Schenker's strategy in interpreting works by Brahms, for example, "is "perhaps not far removed from the work of Hermann Kretzschmar, whose narrative accounts of music's unfolding would later provoke Schenker's notorious scorn" (101).

63 See for example, Gritten and King, eds., *Music and Gesture*. Robert S. Hatten argues that thematic gestures are "basic-level carriers of emotional force" ("A Theory of Musical Gesture and Its Application to Beethoven and Schubert," in *Music and Gesture*, 8). David Lidov provides an example of how, when asked to signal revenge by shaking their firsts in the air, members of different audiences respond unanimously on hearing the short excerpt from the introduction to the trio in Verdi's *Un Ballo in Maschera* to which Lidov's mother had also reacted ("Emotive Gestures in Music and its Contraries," in *Music and Gesture*, 26). And Arnie Cox draws on a concept of mimetic participation to account for embodied responses to music, such as tapping a toe in time to a rhythm ("Hearing, Feeling, Grasping Gesture," in *Music and Gesture*; see also Arnie Cox, "Embodying Music: Principles of the Mimetic Hypothesis," *MTO: A Journal of the Society for Music Theory* 17, no. 2 [2011], accessed November 9, 2011, www.mtosmt.org/issues/mto.11.17.2/mto.11.17.2.cox.html).

64 See Dietrich Bartel, *Musica Poetica: Musical-Rhetorical Figures in German Baroque Music* (Lincoln, NE: University of Nebraska Press, 1997). According to Bartel, the term *anabasis* was first mentioned by Athanasius Kircher. Kircher linked *anabasis* with

> rhetorical figures by naming it with a Greek term. Much more than being simple word painting, the *anabasis* is used to musically recreate the effect of an ascending image or thought found in the text. It is therefore also useful in arousing "exalted, high, and eminent" affections (Kircher).... The figure is both descriptive and affective, describing the text and arousing the

> corresponding affection. Like so many of the musical-rhetorical figures, the *anabasis* can be both image and source of the affection.
>
> (179)

65 Hermann Kretzschmar, "Anregung zur Förderung musikalischer Hermeneutik (1902) ['Suggestions for the Furtherance of Musical Hermeneutics']," in *Musical Aesthetics: A Historical Reader*, vol. 3 of *Aesthetics in Music: The Twentieth Century*, ed. Edward A. Lippman (Stuyvesant, NY: Pendragon Press, 1990 [1902]), 9. Music for Kretzschmar has "no organ for names and designations" (9). Hence the goal of his musical hermeneutics is

> to penetrate the sense and ideational content that the forms enclose, to search out everywhere the soul under the body, to reveal in every part of a work of art the pure kernel of thought, [and] to explain and interpret the whole from the clearest knowledge of the smallest particulars with the use of every aid that technical trading, general cultivation and personal endowment make available.
>
> (6)

Musical hermeneutics consequently came to be regarded as an indispensable aid to understanding an art that stood in need of a narrativizing supplement. See Hermann Kretzschmar, "Neue Anregungen zur Förderung musikalischer Hermeneutik: Satzästhetik (1905) ['New Suggestions for the Furtherance of Musical Hermeneutics: The Aesthetics of Musical Compositions']," in *Musical Aesthetics: A Historical Reader*, vol. 3 of *Aesthetics in Music: The Twentieth Century*, ed. Edward A. Lippman (Stuyvesant, NY: Pendragon Press, 1990 [1905]).

66 See Gary Tomlinson, *Music in Renaissance Magic: Toward a Historiography of Others* (Chicago, IL: University of Chicago Press, 1993). Tomlinson points out that in the order of occult thought's magical episteme, correspondences between musical modes, humors, planetary motions, and the celestial harmony of the heavenly spheres were believed to be the source of music's power. See Chua, *Absolute Music*. Chua similarly maintains that the resulting epistemological shift positioned instrumental music in such a way that its "mechanics of composition (thorough-bass)" (62) required a theory of musical rhetoric to legitimize it. See also Mark Evan Bonds, *Wordless Rhetoric: Musical Form and the Metaphor of Oration* (Cambridge, MA: Harvard University Press, 1991). According to Bonds, the "idea of music as a rhetorical art rests on the metaphor of music as a language" (61).

67 Paul Ricoeur, *The Rule of Metaphor: Multi-Disciplinary Studies of the Creation of Meaning in Language*, trans. Robert Czerny et al. (Toronto: University of Toronto Press, 1977), 284. For a lengthier critique of the role played by this concept of language in music criticism, see my discussions of Charles Seeger's and Lawrence Kramer's treatments of the linguocentric predicament in my *Hermeneutics and Music Criticism*. See also Chapin and Clark, eds., *Speaking of Music: Addressing the Sonorous*; Marie Thompson and Ian Biddle, *Sound, Music, Affect: Theorizing Sonic Experience* (New York: Bloomsbury, 2013). In their introduction, Thompson and Biddle critique musicology's investment in classical social theory as a "silencing of the disturbing turbulences that follow in the wake of Baroque affect theory" (14). According to them, the resurgent interest in sound's affective qualities marks a turn toward aspects of sonic experiences excluded from semiotic or semantic interpretations.

68 Like theories of expressive interpretation, cognitivist accounts that identify music's emotive value with its structural schemata replicate the analytic distinction between music's connotative significance and denotative meanings. This distinction between connotation and denotation corresponds to

a comparable distinction within the tradition of logical positivism, where the difference between implicit meanings and explicit meanings redoubles the difference between cognitive language and emotive language. Ricoeur points out that "a good part of the literary criticism influenced by this positivist tradition transposed the distinction between cognitive and emotive language into the vocabulary of denotation and connotation. For such a position only the denotation is cognitive and, as such, is of a semantic order. A connotation is extra-semantic because it consists of the weaving together of emotive evocations, which lack cognitive value" (Paul Ricoeur, *Interpretation Theory: Discourse and the Surplus of Meaning* [Fort Worth, TX: Texas Christian University Press, 1976], 46). See Anthony Newcomb, "Sound and Feeling," *Critical Inquiry* 10, no. 4 (1984); see also my *Hermeneutics and Music Criticism.*

69 See Chapter 7.

70 Paul Ricoeur et al., *Critique and Conviction: Conversations with François Azouvi and Marc de Launay*, trans. Kathleen Blamey (New York: Columbia University Press, 1998), 174.

71 For a discussion of music's significance with regard to states of mind and our attunement to the world, see my *Hermeneutics and Music Criticism.*

72 I examine music's power to move the heart more closely in Chapter 3; see also my "Feeling, Interiority and the Musical Body," in *Ricoeur and the Body*, ed. Gonçalo Marcelo and Cyndie Sautereau (Lanham, MD: Lexington Books, forthcoming).

73 Carolyn Abbate, *Unsung Voices: Opera and Musical Narrative in the Nineteenth Century* (Princeton, NJ: Princeton University Press, 1991), 53.

74 Paul Ricoeur, *Time and Narrative*, vol. 2, trans. Kathleen McLaughlin and David Pellauer (Chicago, IL: University of Chicago Press, 1985), 78. Ricoeur points out that utterances refer "to the individual who is narrating. Narratology … strives to record only the marks of narration found in the text" (82). The preterite tense consequently marks the entry into narrative. Accordingly, a narrative voice "determines a present—the present of narration—which is just as fictive as the instance of discourse constituting the narrative utterance" (98). In Abbate's view, the lack of a preterite tense in music distinguishes music's mimetic impulses from narrative's diegetic function by precluding the former's capacity for recounting actions and events (Abbate, *Unsung Voices*, 53). See Jeremy S. Begbie, *Theology, Music and Time* (Cambridge: Cambridge University Press, 2000). Begbie attributes disputes over the status of tenses in music to the "myth of [music's temporal] passage" (63). See also Mark Currie, *About Time: Narrative, Fiction and the Philosophy of Time* (Edinburgh: Edinburgh University Press, 2007). Currie's explorations into the significance that narrative's temporal features have for a philosophical inquiry into the nature of time highlight the reflexive relation between experiences of time as they are expressed and articulated in works of art and philosophical speculations. Currie accordingly sets out a tense-based theory as a framework for analyzing the temporal structures and references in the narratives that he discusses.

75 Paul Ricoeur, *Time and Narrative*, vol. 1, trans. Kathleen McLaughlin and David Pellauer (Chicago, IL: University of Chicago Press, 1984), 53. For Ricoeur, "time becomes human to the extent that it is articulated through a narrative mode, and narrative attains its full meaning when it becomes a condition of temporal existence" (52; original in italics); see Ricoeur, *Time and Narrative*, vol. 2. Ricoeur points out that the time of the novel "cannot help but be configured in terms of new norms of temporal organization that are still perceived as temporal by the reader" (25).

2 The myth of Syrinx

Ovid's tale of Syrinx's flight from the lustful Pan affords a glimpse into the mythic origins of music's sublimating power:

> A nymph of late there was
> Whose heav'nly form her fellows did surpass.
> The pride and joy of fair *Arcadia*'s plains,
> Belov'd by deities, ador'd by swains:
> Syrinx her name, by *Sylvans* oft pursu'd,
> As oft she did the lustful Gods elude
> The rural, and the woodland Pow'rs disdain'd;....
> Now while the lustful God [Pan], with speedy pace,
> Just thought to strain her in a strict embrace,
> He fill'd his arms with reeds, new rising on the place.
> And while he sighs, his ill success to find,
> The tender canes were shaken by the wind;
> And breath'd a mournful air, unheard before;
> That much surprizing *Pan*, yet pleas'd him more.[1]

Fashioned from the hollow reeds into which Syrinx was transformed by her water sisters, the music emanating from the pipes that bear her name takes the place of the absent object of Pan's desire:

> Admiring this new musick, Thou, he said,
> Who canst not be the partner of my bed,
> At least shall be the confort of my mind:
> And often, often to my lips be joyn'd.
> He form'd the reeds, proportion'd as they are,
> Unequal in their length, and wax'd with care,
> They still retain the name of his ungrateful fair.[2]

In the words of the Greek poet Nonnus, the music of the satyr's pipes was recompense for the tree-nymph's "too-great pride and love for virginity."[3] Escaping Pan's marriage, Syrinx left the satyr without a bride. Reverberating first with breath of the wind and then with Pan's sighs, the plaintive sound issuing from the hollow reeds—the symbol of the wood nymph's transformation—calls out to the missing Syrinx, so that "she still sings Pan's desire!"[4]

This mythic account of the origin of the sound that gives voice to Pan's complaint provides a unique vantage point from which to begin exploring the reach of music's affective significance. The witching tones that unite the satyr with the disappeared Syrinx are a testament to music's sublimating power. Stripped of its etiological intention, the story of Syrinx's elusive escape, and of the music in which her presence resounds, gives rise to a reflection on music's expressive force.[5]

When in the Introduction I suggested that music becomes profoundly human through staking out the difference between time and its other, I wanted to set the stage for relating music's affective significance to the existential deepening of the experience of time. As I will explain in this chapter, the deepening of this experience gives rise to the enigma of time's ultimate inscrutability. Earlier, I remarked that we can hear the mythopoetic echo of music's sublimation of the deficiency that strikes at the heart of our experiences of time. If, in the myth of Syrinx the music of the pipes reconciles Pan to his loss, this music, which originates with Syrinx's metamorphosis as she passes beyond Pan's grasp, is the symbol of music's power in answer to the pathos of our mortally finite condition.

Music and myth

The myth of Syrinx provides a striking indication of music's expressive force. In the previous chapter, I set out the reasons why music's affective attributes elude the kind of analytic abstractions that identify music's temporal features with the ordinary representation of time. By drawing a configuration from successive elements, the operation that produces these attributes by schematizing them puts into play two competing concepts of time. This chapter in part augments my critique of music's temporal nature by laying out the aporias to which the confrontation between cosmological time and phenomenological time gives rise. In turn, these aporias constitute the context for my claim that music revalues time in response to its deficiency vis-à-vis the other of time. This deficiency, which the contrast between time and eternity brings to the fore, is an inescapable feature of the way that we experience time as the stretching along of our lives. Time for us consists in more than just a series of pinpointed instances that endlessly replace one another in succession. For us, the present is pregnant with the past and the future. In this respect, music both reflects and resolves the aporias arising from the confrontation between cosmological time and phenomenological time through drawing together successively sounding elements.

By taking the myth of Syrinx as my entry point into the broader philosophical field of time, I want in part to set out the difference between the existential deepening of the experience of time and time's erasure. It is difficult to imagine that time would have any meaning at all if it were not for the fact that we reckon with it. Moreover, what interest would we have in eternity if eternity did constitute a limit with regard both to our thinking about time and to our experiences of it? In comparison with our mortal condition, eternity is the emblem of time's other. When in Chapter 6 I distinguish between different representations of the "end of time" in some twentieth-century works and the mythic return to a time *in illo tempore*, I stress that neither these representations nor this return to the *illud tempus* abolishes time. On the contrary, representations of the "end of time" refigure time's ultimate unrepresentability in answer to the enigma of the nonbeing of time. Conversely, the ritual repetition of cosmogenic acts undoes the work of time through restoring the cosmos to its sacred state *ab origine*.

The cultural value that Claude Lévi-Strauss assigns to music and myth underscores the way that he regards both to be instruments of time's obliteration. According to him, the act of listening to music transmutes the segment of time to which this act of attention is devoted into a synchronic totality. This synchronic totality is enclosed within itself, thereby immobilizing passing time. Hence, while listening to music we "enter into a kind of immortality"[6] that is comparable to the way that myth overcomes the contradiction between historical time and its structural constants. According to Lévi-Strauss, music and myth not only transcend the articulate expressions of profane speech; the language of music, which Lévi-Strauss insists is the only one that is both intelligible and untranslatable, is also comparable to the language of the gods. The musical creator thus has a share in the mythic gods' mysterious powers as recounted in myth. From this standpoint, music is therefore also "the supreme mystery of the science of man, a mystery that all various disciplines come up against and which holds the key to their progress."[7]

Attributing music's affinity with myth to their mutual obliteration of time, however, masks how music and myth search out regions of experience that supersede our ordinary ways of reckoning with time. For one thing, the synchronic totality that Lévi-Strauss identifies with a segment of time's transmutation lies on a different axis than the succession of incidents and events. By taking the place of the temporal configuration to which I attributed music's affective force, this synchronic totality paradoxically preserves the system of oppositions and co-ordinations that sets a vertical, synchronic axis against a horizontal, diachronic one. Ricoeur rightly critiques Lévi-Strauss's structural anthropology as "a Kantianism without a transcendental subject,"[8] which Lévi-Strauss points out is the inevitable philosophical consequence of his ethnographic approach. Moreover, Ricoeur stresses that eliminating myth's significance with respect to the "*aporias* of existence ... would be to reduce the theory of myth to the necrology of the meaningless discourses of mankind."[9] By placing myth's function as a narrative of origins in brackets, structural analysis uncovers the systems of relations and oppositions that ostensibly vest myth with its logical function.

Yet myth functions as a logical operation only by combining propositions that point toward boundary situations that myths of origins and ends render meaningful. Myth's logical function is predicated on the narrative resolution of the existential conflicts that give rise to the contradictions that this logical function is said to solve. Accordingly, structural analysis raises the question of myth's—and music's—existential significance to a higher level of radicality.

Music's proximity to myth consequently takes root in the ways that each explores the boundaries between the being and the nonbeing of time. Liminal experiences that revoke the profane order of time's relentless march seemingly share sacred features of the mythic return to a time *in illo tempore*. This return to the "holy time of the beginnings"[10] vests the ritual reenactment of cosmogenic acts with their sacred power. Hence, according to Mircea Eliade, the repetition of events that took place in the Great Time of the ancestors or of the gods overcomes the ritual fault that separates this primal order from the meaninglessness of profane existence. The reenactment of archetypal gestures restores the mythic order of the cosmos through returning to the sacred time of myth. Driven by the "thirst for being,"[11] the ritual repetition of cosmogenic acts is the spring of the eternal recurrence of the same.

Numerous musical ethnographies attest to music's ritual role in trance, ecstasy, and possession experiences. In his ethnography of Tumbuka healing practices in northern Malawi, for example, Steven Friedson identifies the therapeutic value of spirit possession with the trance induced by drumming. In the "thick time"[12] of this possession experience, *vimbuza* drumming calls out the spirits that are the cause of the patient's malady in order to cool them down. The musical heat of *vimbuza* drumming has alchemical-like abilities to transform phenomenal reality into the super-real world of the "dancing prophets."[13] Just as the time of myth differs from that of history, spirit possession induced by *vimbuza* drumming attests to one of the ways that hyperreal worlds of trance and musical ecstasy set themselves apart from the ordinary time of everyday experiences.

The idea that music revalues time in response to the enigma of time's inscrutability marks a decisive break with myth. Freed from its explanatory pretensions, myth's "power of discovering and revealing the bond between man and what he considers sacred"[14] acquires its symbolic significance as a dimension of contemporary hermeneutical thought. Similarly, music's mythopoetic character reveals its existential meaning and significance against the backdrop of time's inscrutable character. My reflections on music's power to refigure time's ultimate unrepresentability are agnostic in this respect. As the proximate analogue of the ritual return to a time *in illo tempore*, music's evocation of a "time beyond time" trades on the hyperreal efficacy of this ritual of return. Yet a striking difference exists between the evocation in music of the feeling of being out of time and the ritual return to a time *in illo tempore*. For, where for mythic thought the return to a time *in illo tempore* is sacred, the aesthetic experiences occasioned by music's explorations of the boundaries between time and its other revalue time within the horizons of our historical situations.

The myth of Syrinx authorizes us to ask further into the ways that, as a mode of human creativity (*poiesis*), music replies to the exigencies of time that prevent the self from occupying the place of its own ultimate foundation.[15] In his reading of the myth, Ernst Bloch points out how the music that resounds in the pipes unites Pan with the nymph who has disappeared yet has not vanished, since her presence is felt in the sound emanating from the hollow reeds. The theurgy that Bloch identifies with this invocation of an essentially human form of expression is the mythic counterpart of music's sublimating power. Through summoning the vanished Syrinx, the "call to that which is missing"[16] resonates with the experience of absence and lack. This absence and lack is the emblem of time's deficiency. More than just the organ's primeval ancestor, the pipes fashioned from the reeds are therefore also the "birthplace of music as a human expression, [which as] a sounding wishful dream"[17] preserves the magical trace of the missing Syrinx. In the myth, "that which has passed beyond the *limit* [of the visible world] is caught up again by this lament."[18] As such, the call of the pipes searches out the absent other in response to the deficiency that places its stamp on our finite, temporal existence.

This mythic account of music's origin thus brings us face to face with a major difficulty for any inquiry into music's temporal features and affective force. The "ontological tonality"[19] that Jean-Luc Nancy searches out at the heart of a form of listening that is resonant with being gives rise to the inconstruable question to which the enigma of time's ultimate unrepresentability also points. According to Nancy, "being continually differs from all being-here-and-now,"[20] so that the living present resounds in the *différant* between being and presence. Hence, the ontological tonality that reverberates with the presence of a sonority that eludes our sight sets the *différant* in the place of an unknown and unknowable order of alterity. This evocation, which for Nancy summons presence to itself, echoes the mythic call of the pipes. Accordingly, this archaic call is both the emblem of the wound inflicted by the difference between temporal being and an order that exceeds it, and the sign of the desire from which the hope for transcendence springs.

Interlude: *Syrinx*

The languorous air of Claude Debussy's *Syrinx* seems at first only remotely related to Ovid's account of the origin of the Arcadian wood god's pipes. Composed in 1913 for Gabriel Mourey's unfinished play *Psyché* (originally called *Flûte de Pan*), Debussy's unaccompanied composition for solo flute evokes the erotic sensuality celebrated by artists and poets of the time. The work's languid expression of a Symbolist idyll echoes the pastoral call of Pan's mythic plaint in its opening line (Example 2.1).

Example 2.1 Claude Debussy, *Syrinx*, mm. 1–2.

An evocative example of the Symbolists' fascination with erotic eclogues populated with naked nymphs and satyrs, *Syrinx* recalls the mythic origin of music's sublimating power through the soporific reverberation in its erotic sensuality of the longing for being. Three years before Stéphane Mallarmé published his poem *L'après-midi d'un faune* (1876), on which Debussy's *Prélude à l'après-midi d'un faune* (1894) was based, William-Adolphe Bouguereau—the "master of the artistic pinup"[21]—exhibited his *Nymphs and Satyr* (1873). According to Mallarmé, the subject of his poem was an ancient symbol, the power of which also resounds in the sensuous tonalities of Debussy's *Syrinx* and *Prélude à l'après-midi d'un faune*.[22] The erotic character of these tonalities accentuates another dimension of the mythic portrayal of the wood nymph's desperate flight. In *Syrinx* and *Prélude à l'après-midi d'un faune* (the sensual atmospherics of which were redoubled by Vaslav Nijinsky's choreography for the 1912 performance by the Russian Ballet), Eros accordingly regains something of the cosmic force that it once had in sacred mythologies in these musical eclogues' atmospheric sensations of restless, sexual desire.[23]

Syrinx's somnolent evocation of this feeling of desire is indicative of an aesthetic rooted in the rejection of the dramatic pathos of Romantic and especially Wagnerian sensibilities. The "principle of ornament"[24] that Debussy espoused celebrated the figurative character of music's sensuously sonorous materials. Often associated with Impressionism (a term that the critic Louis Leroy used in a derogatory reference to Claude Monet's *Impressions: Sunrise*), Debussy's music has as much if not more in common with Symbolist ideals. (Like Symbolist poets who regarded a poem as a self-signifying gesture and who used words for their musical qualities, music for Debussy, who frequented Mallarmé's Symbolist gatherings, was an "abstract ballad"[25] fashioned from the sum of its variously scattered forces.)

Syrinx accordingly conjures its Arcadian root through summoning the latter's erotic atmosphere. The languid return of the opening gesture (Example 2.2) accentuates this charged atmosphere's heightened sensuality. It is as though, in the absence of any visual representation of the missing Syrinx, the agogic accents of the returning b♭s, together with the chromatic flourish (d♭-gf♭-e♭♭-d♭-b♭) followed by the languorous descent to b♭, gives voice to the torpid sexuality of Pan's complaint. This muted expression of restless desire echoes the archaic call to the missing Syrinx. The sentiment that in Debussy's *Syrinx* assumes the guise of a Symbolic idyll is thus redolent of the call that resounds in the hollow space of the panpipe's reeds.

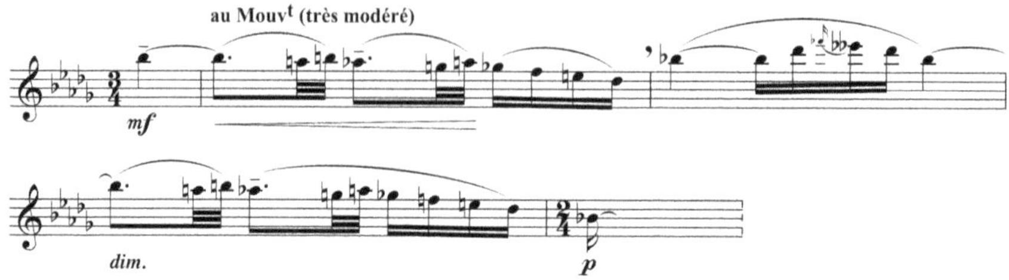

Example 2.2 Claude Debussy, *Syrinx*, mm. 26–28.

Species of musical time

Jonathan Kramer's typology of species of musical time represents one of the most systematic attempts to account for the differences between the ordinary concept of time and the variety of ways that music evokes a deepening sense of eternity. As critical of the excesses of formalistic practices as he is of a supposedly humanistic orientation that favors purely subjective impressions over more rigorous modes of inquiry, Kramer regards music's temporal constitution to be fundamental to its meaning and significance. Music is consequently not only "meaningful in and through time."[26] Time for Kramer is also "both the essential component of musical meaning and the vehicle by which music makes its deepest contact with the human spirit."[27]

The five species of musical time identified by Kramer comprise a hierarchical order that sets the representation of temporal movement against that of a static "standing present" at the cost of eclipsing the synthetic operation that, as I explained, draws expressive figures from a succession of tones. The first species, which Kramer argues predominates in Western art music, is that of directed linear time. Tonal motion, which Kramer admits is strictly speaking a metaphor, embodies the teleological sense of time that he maintains is common to our Western mentality and life. Standard tonal progressions evince the kind of goal-directed movement that is the paradigmatic feature of directed linear time. For Kramer, the regulated succession of tonal harmonies constitutes forward-directed temporal movement (the sense of which, in the previous chapter, I attributed to the operation that draws a configuration from tones, harmonies, and the like that follow one another). This first species of time, which Kramer tells us is expressed quintessentially by the tonal system, is thus the approximate equivalent analogue of the ordinary concept of time.

The second species—nondirected linear time—is a variant of the first. Here, teleological expectations give way to those of textural or rhythmic consistency. When in listening to a work we become aware of a rhythmic or textual pattern's invariant quality, nonlinear features take on an increasingly significant role. The recurring arpeggiation in Bach's "Prelude in C Major" from *The Well-Tempered Clavier* (Book 1; Example 2.3) is a case in point. The regularly patterned texture, together with its sustained, interlocking suspension figures, evinces the lack or absence of a "clearly implied goal, despite the directed continuity of motion"[28] that for Kramer is a dominant feature of this second species of time. The ostensibly nondirected character of recurring arpeggiated figures stands out against the directed harmonic progression to which these unfolding figures give a body. As such, the representation of nondirected linear time in this passage takes as its reference point the regulated succession of tones and harmonies within a goal-oriented framework.

Example 2.3 J. S. Bach, "Prelude in C Major" from *The Well-Tempered Clavier* Vol. 1, mm. 1–4.

The third species—multiply-directed time—stands midway between linear time and its corollary opposite, vertical time. While for this species of time the "sense of goal-direction is [still] acute,"[29] the expected goal or goals are equivocal. The representation of multiply-directed time in tonal music relies heavily on the teleological inferences of gestural conventions such as beginnings, cadences, musical climaxes, etc. Suppressing or thwarting these conventions' implied goals ostensibly subverts their linear implications. Kramer cites as an example the idiosyncratic placement of closing and opening musical gestures in Beethoven's *String Quartet* in F Major, op. 135, which he tells us disrupts normative expectations. When in the tenth measure of the first movement "we hear the profile of the final cadence,"[30] absolute time (by which Kramer means time as objectively ordered by the fixed linear succession of beats or seconds, for example) is effectively reordered. In most tonal music, Kramer maintains, "we still hear agreement between the formal implications of absolute time and those of gestural time."[31] Multi-directed linear time thus breaks up the regulated order of linearly directed time, where the measured progression of successive tones, harmonies, timbres, and the like conforms to absolute's time's quantifiable nature.

The distinction Kramer draws between absolute time and gestural time highlights a major weakness in his theory. Kramer's assertion that "time does not exist apart from experience"[32] seemingly anchors his typology in a phenomenological approach. Yet the various temporalities that he identifies with the different species of time are predicated on time's ordinary representation as a series of instantaneous "now" points. According to Kramer, the appearances of the closing cadential gesture (mm. 5–10, 104–109, and 188–193) in the first movement of Beethoven's Opus 135 call into question the "very meaning of musical continuity, and it directly postulates a multiple temporal continuum."[33] Hence, on his account, discrepancies between the formal implications of contextually defined gestures and their sequential ordering give rise to a time experience that deconstructs traditional and even nostalgic representations of time. But how, we might ask, could we grasp this cadential gesture's temporal incongruity apart from hearing how it figures within the larger passage that provides its musical context? The difference that Kramer insists exists "between time as used (absolute time, governed by the inevitability of succession and the syntax of tonal progression) and time as portrayed (gestural time, as suggested by the inherent temporal functions of gestures)"[34] also underscores the connection between music's sequential dimension (where tones follow one another in succession) and its temporal configuration. Consequently, the significance that Kramer attributes to the discrepancy between a gesture's formal implication and its contextual placement rests on the synthetic operation that draws together temporally dispersed elements into meaningful configurations.

The fourth and fifth species further underscore the difficulty of accounting for music's temporal constitution independently of the phenomenology of time. If, on the one hand, these two species of time approximate time's existential deepening (which I pursue in the next section in connection with the confrontation between the phenomenology of time and the cosmological concept of time), on the other hand, these species' reliance on time's ordinary representation beggars the hierarchical order of his typology. In Kramer's system, moment time and vertical time (the fourth and fifth species, respectively) stand at the opposite pole to the concept of time that Kramer identifies most closely with Western cultural concepts and practices. As such, these species of nonlinear time are representative of the way that time as the medium of movement and change gives way to its other.

Moment time is a case in point. Here, the sense of movement that Kramer identifies with tonal music's teleological structure is permanently arrested. Moment form "verticalizes one's sense of time within sections"[35] so that each moment is related to others in a work by means of a nonlinear logic. For Kramer, this logic (which he identifies with the structure of a moment form work such as Karlheinz Stockhausen's *Kontakte* [1960]) is comparable to the "seemingly irrational"[36] temporality of the mind. However, it is again difficult to see how a succession of isolated moments could contribute to their cumulative effect independently of the operation that draws them together. Kramer is right to differentiate between the logic constituted by the succession of isolated moments and the logic of a tonal work. Yet, as I pointed out earlier, this logic is also drawn from the way that isolated moments contribute to structuring their temporal succession. Moment time's radical disruption of linearly directed time's teleological orientation clearly subverts the latter. Yet, apart from the synthetic apprehension of moments that follow one another in succession, the sense of structured portions that are the telltale feature of this species of time would dissolve into a series of temporally extended "now" points (each of which redoubles the paradox in question).

The difficulty that inheres in Kramer's categorization of the species of time reaches its highest point of virulence with the fifth species. By setting the synchronic dimension of time's "verticalization" against the diachronic axis of temporal changes, Kramer vests this species with its radical significance at the same time that he binds it to the ordinary concept of time. According to him, in this species, "discontinuity disappears in favor of total, possibly unchanging, consistency."[37] The "timeless now of the extended present"[38] thus sets the eternity of the moment against time's inexorable advance. It is as if "a single present [is] stretched out into an enormous duration, a potentially infinite 'now' that nonetheless feels like an instant."[39] We could agree that time in its ordinary sense disappears, *as if* "frozen in an eternal now."[40] This "time of timelessness,"[41] as I will explain further in Chapter 6, is accordingly the simulacrum of a time beyond time that allows time to be surpassed by its other. Although Kramer cautions against drawing correlations between his hierarchical ordering of the species of musical time and J. T. Fraser's theory of time's evolution, it is impossible to overlook how time's supposed evolutionary stages line up with Kramer's categorizations of the species of musical time.[42] The more primitive levels of these evolutionary stages are indicative of unchangeable and eternal states of *being* that Kramer tells us are more closely related to Eastern cultures. Conversely, the higher levels of these evolutionary stages are more readily identifiable with Western cultural practices and traditions, since they are indices of states of *becoming*. For Kramer, the evocation of a "time of timelessness" puts us in contact "with a deeply human time sense that is often denied in daily living (at least in Western cultures)."[43] In contrast, from the hermeneutical vantage point that I have set out, music's mimetic expressions of being beyond or out of time sound out the fundament in giving a figure to the other of time's architemporal order.

Phenomenological time and cosmological time

Augustine's question, "What, then, is time?"[44] inaugurates a different type of discourse. By placing his meditation on time against the backdrop of eternity, Augustine brings the aporetic character of the experience of time to the fore. Unlike God's eternal present, the present for us is distended and dispersed across the past and the future. Hence, Augustine considers that in reciting *Deus creator omnium* (the psalm he uses as his example), the faculties of attention, memory, and expectation are all in play:

> Suppose I am going to recite a psalm that I know. Before I begin my faculty of expectation is engaged by the whole of it. But once I have begun, as much of the psalm as I have removed from the province of expectation and relegated to the past now engages my memory, and the scope of the action which I am performing is divided between the two faculties of memory and expectation.... But my faculty of attention is present all the while, and through it passes what was the future in the process of becoming the past.[45]

Augustine's inestimable discovery was to tie the distention of the soul to the slippage between the present of the future, the present of the past, and the present of the present. As Ricoeur explains, only a mind (soul) that distends itself across this threefold present can engage itself. It is only "*in* the soul, hence as an impression, that expectation and memory possess extension."[46] Attention is therefore also an intention, insofar as the "transit through the present ... become[s] an active transition."[47] The soul's distention (*distentio*) springs from the "noncoincidence of the three modalities"[48] of the action of attention, memory, and expectation. Time's extension corresponds to the soul's distention (*distentio animi*). Since the present as we experience it is not a pinpointed instant, the *distentio animi*'s threefold intention (tied to the faculties of attention, memory, and expectation) thus actively bursts asunder a present intention (*praesens intentio*).

By linking the measurement of time to that of a present conceived in terms of the faculties of attention, memory, and expectation, Augustine's solution to his question concerning the being of time gives rise to a series of enigmas. First, the measurement of a long or short time with respect to memory and expectation lacks the support of spatial reference points that mark the trajectory of something in motion. Second, there is no "independent mode of access ... to the extension of the impression [of the length of time] inasmuch as it is held to be purely 'in' the mind."[49] Third, the metaphor of transit appears to be the only way of expressing the connection between the soul's intention (*intentio*) and its being-affected (*affectio*) with regard to the spaces that expectation, attention, and memory traverse. Finally, the "soul 'distends' itself as it 'engages' itself."[50] This, for Ricoeur, is the supreme enigma. By tying the distention of the soul to the slippage that finds its way into the threefold present, Augustine's solution to the aporia of time's measurement highlights the discordance that repeatedly emerges from the concordance of the soul's threefold intentions. In sum, the "more the mind makes itself *intentio*, the more it suffers *distentio*"[51] as a result of its own activity.

By contrasting time with eternity, Augustine sets the soul's experience in relief. Through pushing the reflection on the *distentio animi* as far as possible, Augustine compares the slippage of the present of the present, the present of the past, and the present of the future to a "standing" present for which there is neither past nor future. The thought of eternity functions as a limiting idea, "against the horizon of which the experience of the *distentio animi* receives, on the ontological level, the negative mark of a lack or a defect in being."[52] In contrast to this experience, the eternal present that declares itself in the first person can only be thought of as "supreme height."[53] An original nothingness—anterior to the universe's creation *ex nihilo*—thus "strikes time with an ontological deficiency"[54] that places its stamp on the difference between time and its other.

Ricoeur's stunning analyses of the aporias engendered by the phenomenology of time draw out the consequences of Augustine's meditations. Not only is the major aporia with which Augustine is struggling "inscribed within the circle of an aporia that is even more fundamental, that of the being or the nonbeing of time,"[55] but the ontological nature of Augustine's question "*quid est enim tempus?* ('What, then, is time?')"[56] also opens an inquiry that leads to time's internal hierarchization.

The thought of eternity impacts speculations on time in at least three ways. First, the thought of eternity highlights time's deficiency by placing speculations on time "within the horizon of a limiting idea that forces us to think at once about time and about what is other than time."[57] Second, the meditation on eternity not only intensifies the experience of the soul's distention on the existential level; more than just a limit for thought, the absence of eternity is also experienced in terms of a "lack that is felt at the heart of [our] temporal existence."[58] Third, by calling upon the experience of *distentio* "to surpass itself by moving in the direction of eternity,"[59] the contrast with eternity reveals an internal hierarchy that deepens the experience of time (see below). The contrast between time's dispersion (which is the cause of the soul's distention and the principle of its [self-] alienation[60]) and the oneness of the eternal present takes root in the symbolism that gives a figure to the "sorrow of the finite and the celebration of the absolute."[61] Music's refiguration of time's ultimate unrepresentability augments the symbolism of this sorrow, which is born from time's deficiency, and of the joy that springs from the sublimation of sadness. In answer to the absence of aseity and to the element of passivity at the heart of our temporal existence, music gives voice to feelings of sorrow and joy through staking out the difference between time and eternity, as I explain in Chapter 3.

By deriving the principle of time's extension and measurement from the mind's distension, Augustine successfully refutes the cosmological thesis relating time to the movement of the heavenly bodies (on which the ordinary concept of time is based), but only at the price of engendering an aporia that is even more intractable than the one to which the soul's distention gives rise. This aporia stems from the fact that there are two conflicting ways of thinking about time: phenomenological and cosmological. The approaches to the problem of time with respect to these two ways of thinking about it are mutually exclusive. Accordingly, Aristotle destroys "Augustine's attempt to found the measurement of time in the distension of the soul"[62] in advance by relating time to movement without equating the two. Unlike Augustine, for Aristotle the perception of time thus depends on the perception of movement so that the mind discovers succession in the external world before constructing it.

The difficulty of conceiving time in terms of the split between the movement of which it is an aspect and the soul that discerns this movement, together with the difficulty of conceiving movement itself, which is neither power nor act, mark out the aporias that undercut Aristotle's arguments. Time, it turns out, can no more be conceived purely in terms of a cosmology than it can be explained by a psychology. Hence, neither Aristotle nor Augustine has the final word. The confrontation between the phenomenology of temporality and cosmology draws speculations on time to the brink of a precipice. For the "conceptually unbridgeable gap between the notion of the 'instant' in Aristotle's sense and that of the 'present' as it is understood by Augustine"[63] destroys any comprehensive approach that would overcome this gap systematically. Time's physical extension can no more be derived from the soul's distention than the soul's distention can be derived from the physical extension of time. The mutual exclusion of cosmological and phenomenological accounts of time thus gives rise to an aporia that can be resolved poetically, and not theoretically or speculatively, through the ways that individual works draw together successive incidents (in the case of narrative) or sounding elements (in the case of music) as meaningful, temporal wholes. (The idea that movement in music consists in the succession of its constitutive elements redoubles the difficulty of attributing music's temporal character either to a cosmological or to a phenomenological concept of time. For, as I explained in Chapter 1, music's temporal character is a feature of the configurating operation that also gives rise to music's affective attributes.)

The confrontation that Ricoeur constructs between Husserl's phenomenology of internal time-consciousness and Kant's thesis regarding the invisibility of time intensifies the aporia brought to light by the confrontation between cosmological and phenomenological concepts of time. By excluding world time, Husserl intended to make lived time itself appear. Yet the consciousness of lived time presupposes the constitution of the enduring unity of a temporal object, which Husserl hoped to derive from the constitution of the pure temporal flow. By attempting to derive objective time from our internal consciousness of time, Husserl inverted the relation of priority between them. Husserl wanted to ensure that objective time's constitution would be independent of the constituting operation internal to time consciousness.[64] Yet the unity of the flow of time internal to consciousness can be expressed only with the support of objective time, which Husserl's phenomenology rescinds in advance.[65]

Kant's thesis—that while time is a condition of phenomenal appearances, time itself does not appear—stands over against Husserl's phenomenological investigations. According to Kant, "[t]ime is the formal *a priori* condition of all appearances whatsoever."[66] Hence, rather than perceive time, we perceive objects in time. Kant accordingly identifies three modes of time: duration, succession, and coexistence. The principle of time's permanence (that is, the principle that time exists independently of changes taking place within it) designates the underlying ground such that two phenomena could be said to be successive or simultaneous. Ricoeur notes that "since time itself cannot be perceived, it is only by way of the relation between what persists and what changes, in the existence of a phenomenon, that we can discern this time that does not pass and in which everything passes."[67] This relation therefore governs the duration of a phenomenon. The regularity of succession and the simultaneity of phenomenal events constitute the basis for the transcendental determination of time as the condition of the events that occur in it. Yet Kant's transcendental approach proves to be no more self-sufficient than does Husserl's phenomenological one. On the contrary, the temporal irreversibility of a succession of events depends upon the connection thematized by the phenomenologist between the present, the past, and the future. Hence, the difficulty: "phenomenology and critical thought borrow from each other only on the condition of mutually excluding each other."[68]

Martin Heidegger's hermeneutic phenomenology intensifies the aporias that haunt Augustine's and Husserl's thought. The question of Being-a-whole with respect to one's own life offers a privileged point of access to the problem of time as governed by the highest level of authenticity and its criterion of resoluteness in the face of one's inevitable death. Heidegger relates the question of time to *Dasein*'s mode of being by tying time's authentic structure to that of Care.[69] The present finds its meaning through the intermediary of the situation that offers its possibilities to and imposes its constraints on us. Accordingly, *Dasein*'s having-been arises from coming-toward itself in accordance with its ownmost possibilities. Temporality, Ricoeur explains, "is then the articulated unity of coming-toward, having-been, and making-present, which are thereby given to be thought of together."[70] By carrying over the Augustinian problem of the threefold present, the temporalization of temporality reintroduces the enigma of the *distentio animi* via the unification and diversification of the three temporal ecstases. Heidegger appeals to temporality's authentic structure (which is the highest level of the three levels) as the guarantee of his analysis's primordial character. Historicality (the second hierarchical level) adds the aspect of the stretching along of a life between birth and death. *Dasein* accordingly "constitutes its true being as this very stretching-along, which envelops its own beginning and its own end, and gives meaning to life as 'between.'"[71] Within-time-ness (the third level) completes the idea of historicity, from which it is also derived, by placing the accent on the force that the present has with regard to the initiatives that we take. While still a feature of Care, reckoning with the time of everyday affairs represents a leveling off of the most authentic level of temporality for Heidegger. However, Ricoeur also points out that Heidegger's quest for authenticity "cannot be carried out without a constant appeal to the testimony of the existentiell."[72] For just as the analysis of temporality leads to historicality's derivation, the structure of within-time-ness completes that of historicity, from which as previously indicated it is also derived.

Far from supplanting the opposition between Augustine's and Aristotle's conceptions of time, and between Husserl's phenomenology and Kant's transcendental deduction, the analytic of *Dasein* also engenders the aporias that are actively at work in it. These aporias reach their highest point of virulence in a phenomenology of time that discovers its external limit when it "reaches those aspects of temporality that are most deeply hidden, even when they are closest to us."[73] The failure to generate time's autonomy with respect to movement (despite the proximity of within-time-ness to the ordinary concept of time) constitutes the phenomenology of time's ultimate aporia. In summing up this "failure of our thinking about time,"[74] this aporia promotes the internal diversification within the existential analysis of temporality, historicality, and within-time-ness. Just as importantly, the interference between an existential and an empirical way of thinking about time not only brings to the fore a feeling of complicity "[b]etween thrownness and fallenness, which constitute our fundamental passivity with regard to time,"[75] and the stars' sempiternal revolution. This fundamental element of passivity also attests to the alterity of time's other through inscribing the absence of aseity at the heart of our finite existence.

Music and the inscrutability of time

The idea that music is most profoundly human when, in response to the temporal condition of mortal life, it explores regions where time is surpassed by its other underscores the value of engaging hermeneutically with the confrontation between cosmological time and phenomenological time. The aporias that spring from these different ways of thinking about time not only accentuate the significance that music has through drawing together elements that sound in succession, but these aporias also bring to the fore music's express value with respect to the pathos of our mortally finite existence. Hence, where philosophical speculations on time give rise to a series of aporias, music's mimetic responses to the difference between the being and the nonbeing of time multiply the ways that time is surpassed by its other.

The inability of the phenomenology of time to engender the order of time that it prescinds—namely, cosmic time—gives rise to another series of aporias, the last of which is the aporia of time's inscrutability. Ricoeur's thesis in *Time and Narrative*—that the poetics of the narrative art reply to the aporias of temporality—underscores the significance of the first aporia. Phenomenological time's and cosmological time's mutual occultation sums up the failure of our thinking with respect to time's constitution when, in Heidegger's analysis, the recovery of time's genuine constitution fails to abolish time's ordinary representation as a linear succession of instantaneous "now" points. In response, narrated time, which the act of emplotting events in the telling of a story produces, is "like a bridge set over the breach speculation constantly opens between phenomenological time and cosmological time."[76] Music, too, offers its specific mediation through evoking feelings and moods. Like narrative, music creates a "third-time"—musical time—as a feature of its temporal configuration.

The second aporia in the series of aporias that spring from the mutual occultation of phenomenological time and cosmological time attests to the impossibility of totalizing time within the sphere of our historical being, or historicality. This aporia springs from the dissociation of the three temporal ecstasies (future, past, and present) "despite the unavoidable notion of time conceived as a collective singular."[77] Ricoeur asks whether thought about history might provide a practical response to this aporia, since phenomenology is unable to offer a theoretical one. In renouncing Hegel's speculative solution along with the hubris of the kind of thinking that dares to elevate itself to the level of the absolute in order to grasp "history as the totalization of time in the eternal present,"[78] Ricoeur ties the aporia of time's totalization with respect to history to the imperfect mediation between the space of our experiences and the horizons of our expectations.[79] The force of the present acquires its ethical and political dimensions only by reason of the initiatives that we take to preserve the tension between the space of our experiences and the horizons of our expectations. Consequently, this imperfect mediation constitutes the practical response to the aporia of the oneness of time.

This aporia has a critical place in my reflections. When in the final chapter I draw out the resemblance between music's power to augment the practical field of our experiences and exemplary moral and political acts, I relate the force of the present to the loss of credibility of the philosophy of history that elevates thought about time to the level of the absolute. My reflections on Schoenberg's renunciation of tonal closure in Chapter 5 in part set the stage for my subsequent remarks on the ethical and political challenges stemming from the fact that there is "no plot of all plots capable of equaling the idea of one humanity and one history."[80] Apart from this idea, the hope that freedom could become a historical reality would be nothing more than a lifeless transcendental.[81] Consequently, the aporia of the oneness of time is a critical touchstone for drawing out the connection between music's refiguration of time's ultimate unrepresentability and the practical task that reason sets itself in carrying out the imperfect mediation between the space of our experiences and the horizons of our expectations, in accordance with the idea of justice's federating force.

The third aporia is the most radical of the three. This aporia surges forth at the point at which the confrontation between cosmological time and phenomenological time is most virulent. When, "at the moment when time, escaping any attempt to constitute it, reveals itself as belonging to a constituted order always already presupposed by the work of constitution,"[82] time's ultimately inscrutable character unseats our effort to think about time. The enigma of time's inscrutability consequently shatters the conceit that compels our thought to proclaim itself to be the master of meaning. Ricoeur emphasizes that, when in Heidegger's analysis the distinction between temporality (*Zeitlichkeit*) and temporal-being (*Temporalität*) is "separated from the idea of the Good,"[83] only the element of the passage beyond remains. By inquiring "even beyond being as to that upon which being itself as being, is projected,"[84] Heidegger's analysis points to a phenomenon that is inaccessible to any hermeneutic phenomenology as such. At the limit of thought, the ontological difference between our temporally finite existence and the alterity of an order onto which our temporal being is projected as its condition of possibility is the emblem of the absence of aseity. In answer to this absence, music revalues time in response to time's ultimate inscrutability.

We might wonder whether the art of narrative is the sole "guardian of time"[85] in this respect. Ricoeur's thesis that narrative discourse mediates the aporias engendered by phenomenological and cosmological perspectives on time runs up against its own limits as the narrative art exhausts itself in attempting to draw near the inscrutable character of time; narrative here flows over into other genres of discourse that in their own ways speak of time and of the other of time. The degree of intoxication that narrative allows itself in multiplying limit-experiences worthy of being placed "under the sign of eternity"[86] has the sobriety of phenomenology as its counterpart. By conferring a narrative transcription on the archaisms and hermeticisms from which phenomenology distances itself, the art of narrative appropriates the substance of these archaisms and hermeticisms in speaking of the other of time. Music goes even further. Through staking out the difference between time and eternity, music explores the variety of ways that boundary experiences revalue time.

The resurgence of the aporia of time's inscrutability consequently figures prominently in my reflections on how different works refigure time's ultimate unrepresentability through searching out the borderlines between time and the other of time. By returning speculations on time to nonspeculative modes of thinking, in which archaic inspirations of the thought of time and the other of time can still be heard, this aporia spans the chasm separating music's historical significance and the power of myth. Like the myth of Syrinx, which beyond its etiological intention speaks of the origin of music's sublimating power, one such inspiration is Greek. Here, the thought of the pre-Socratics Parmenides, Heraclitus, and Anaximander "points toward a region where the claim of a transcendental subject (in whatever form) to constitute meaning no longer holds sway."[87] Ricoeur explains that by dwelling "alongside an *arkhe* that is the condition of possibility for all the presuppositions we can posit,"[88] this pre-Socratic thought repeats the archaism of a mythical beginning with which it also breaks. Only myth's philosophical retrieval can bring the genesis of time to language, as Plato's philosophical story *Timaeus* attests. Conversely, only a "form of thinking that renders itself archaic can understand Anaximander's saying, whose voice can still be heard … as the isolated witness to this time that remains inscrutable as much for phenomenology as for its other, cosmology."[89] The mark of the inscrutable that these mythic archaisms place on speculations on time—together with the evaluations of human time that take place across these polymorphic representations of something beyond time—provide critical reference points for drawing out the force and power of music's refiguration of time's ultimate unrepresentability. Mythic figures that assimilate time to a circle abound, extending from the cyclic organization of the seasons and years to the eternally immobile identity of the divine *aion*, "which, detached from this image of a circle, …is then connected with the unending round of births and rebirths."[90] The idea that all things derive their existence from the source to which they necessarily return "according to the arrangement of Time"[91] is one ground of some of the unavoidable figures of time's ultimately inscrutable character.

The second archaic inspiration is Hebraic. The contrast between time and eternity that envelops Augustine's examination of time preserves the trace of this Hebraic archaism. Contrasting a present without future or past to a transitory experience highlights the deficiency of time in relation to the fullness of *nunc stans*. Here, our temporal existence is marked by its condition of lack in relation to the plenitude of eternity. Eternity's anteriority with respect to time accentuates the contrast between the being and the nonbeing of time; as a limiting idea under which we can think about time, eternity's anteriority with respect to time is also the emblem of the other of time's supreme height. Here again, the diversity of figures in music of something beyond time calls for a reflection on music's sublimating power. The tonalities of lamentation, hope, and praise that accompany Augustine's speculation on time and eternity thus not only resonate with this Hebraic inspiration, but they also have a place in my studies of the ways that music replies to the deficiency brought to the fore by the contrast between time and eternity. This second archaic inspiration stands over against eternity's reduction to an immutable present. Stories of beginnings such as those recounted in Genesis and Exodus "speak of eternity inasmuch as a certain fidelity is founded in them."[92] By opposing "the eternity of God to the transitory character of human life,"[93] hymns and wisdom literature that culminate in songs of lamentation and praise give voice both to the sorrows and joys of mortal beings who are subject to time's wearing effects. The variety of the tonalities of these stories, hymns, and lamentations therefore accords with "an essentially nonspeculative, nonphilosophical mode of thinking, for which eternity transcends history from within history."[94] As such, these tonalities are a sign and pledge of the promise of hope.

The notion that music refigures time's ultimate unrepresentability in answer to time's inscrutable character marks a decisive turning point in my analysis. Through searching out the boundaries between time and its other, music's figurations of a "time beyond time" preserve traces of these archaic Greek and Hebraic inspirations. The reverberation of a mythic return to a time *in illo tempore* or of eschatological representations of the "end of time" in music thus draw my reflections in two directions. The difference between music's approximation of this mythic return and of eschatological presentiments will prove to be crucial when it comes to considering music's impact on the ways in which music opens the world to us in response to the difference between time and its other. It will be the task and challenge of the next chapters to explore different ways in which music evokes sentiments, feelings, and moods in answer to the enigma of the being and the nonbeing of time.

Notes

1 Ovid, *Metamorphoses* [689–695; 704–709], trans. Sir Samuel Garth et al. (Hertfordshire: Wordsworth Editions, [1826] 1998), 30–31.
2 Ovid, *Metamorphoses* [710–716], 31.
3 *Nonnus Nonnos of Panopolis, Nonnos Dionysiaca,* vol. 3, bk. XLII, trans. W. H. D. Rouse (Cambridge, MA: Harvard University Press, 1940), 255.
4 *Nonnus, Nonnos Dionysiaca,* vol. 3, bk. XLII, 255.
5 See Paul Ricoeur, *The Symbolism of Evil*, trans. Emerson Buchanan (Boston, MA: Beacon Press, 1967). Beyond its demythologization, myth becomes a "dimension of modern thought" (5). For us, a "myth is *only* a myth because we can no longer connect that time with the time of history as we write it, employing the critical method, nor can we connect mythical places with our geographical space" (5).
6 Claude Lévi-Strauss, *The Raw and the Cooked: Introduction to a Science of Mythology*, vol. 1, trans. John and Doreen Weightman (Chicago, IL: University of Chicago Press, 1969), 16. According to Lévi-Strauss, the way that the spaces set off by the sacred stand apart from the regions of profane discourse is comparable to the way that "singing or chanting differs from the spoken language" (28).
7 Lévi-Strauss, *The Raw and the Cooked*, 18.
8 Paul Ricoeur, *The Conflict of Interpretations*, ed. Don Ihde (Evanston, IL: Northwestern University Press, 1974), 52; see Claude Lévi-Strauss, *The Raw and the Cooked*, 11.
9 Paul Ricoeur, *Hermeneutics and the Human Sciences: Essays on Language, Action and Interpretation*, trans. John B. Thompson (Cambridge: Cambridge University Press, 1981), 161.
10 Mircea Eliade, *Myths, Dreams and Mysteries: The Encounter between Contemporary Faiths and Archaic Realities*, trans. Philip Mairet (New York: Harper and Row, 1960), 23. The return to a sacred time *in illo tempore* through these ritual reenactments consecrates the real through the repetition of cosmogenic acts. Eliade explains that myth "is thought to express the *absolute truth*, because it narrates a *sacred history*; that is, a transhuman revelation which took place at the dawn of the Great Time." As a "*true history* of what came to pass at the beginning of Time," myth thus provides the pattern for human behavior. See Mircea Eliade, *The Myth of the Eternal Return: or, Cosmos and History*, trans.

Willard R. Trask (Princeton, NJ: Princeton University Press, 1954). See also Ricoeur, *The Symbolism of Evil*, 5.

11 Eliade, *Myth of the Eternal Return*, 11.

12 Steven M. Friedson, *Dancing Prophets: Musical Experience in Tumbuka Healing* (Chicago, IL: University of Chicago Press, 1996), 124. Friedson explains that a

> theory of correspondence relates spiritual energy to rhythmic modes. Only the rhythmic mode of *vyanusi* will heat up the spirits of the Ngoni and transform ordinarily shy young women into spear-welding Ngoni warriors. Similarly, only the rhythms of *fumu za pasi*, with its distinctive combination of open and stopped strokes will call for the *mphepo* (spirit-wind) of the lion, which brings the threat of violence but also the promise of great divinatory power.
>
> (129)

See also Chapter 6 in this book.

13 Friedson, *Dancing Prophets*. In the "act of dancing *vimbuza* … the ancestral *mizimu* and foreign *vimbuza* are 'melded together' (the term the Tumbuka use), creating the necessary conditions for an *nchini* to 'see' (*kuwona*)" (22).

14 Ricoeur, *Symbolism of Evil*, 5.

15 See my "Introduction: Paul Ricoeur and the Age of Hermeneutical Reason," in *Paul Ricoeur in the Age of Hermeneutical Reason: Poetics, Praxis, and Critique*, ed. Roger W. H. Savage (Lanham, MD: Lexington Books, 2015).

16 Ernst Bloch, *Principle of Hope*, vol. 3, trans. Neville Plaice et al. (Oxford: Basil Blackwell, 1986), 1059. Original in italics. According to Bloch, it is impossible to avoid hearing such a call in singing. For this call, which began first as a cry, "expresses a drive and calms it at the same time" (1058).

17 Bloch, *Principle of Hope*, vol. 3, 1059. "The sound of the shepherd's pipe, the panpipe, or of the syrinx as the Greeks called it, … is supposed to reach the distant beloved." On Bloch's account, the panpipe originated among a "different social class than [that of] the noise-producing, frighteningly cultic musical instruments …. like the twirling stick, cymbals, or the magically painted, itself magically worshiped drum." By entering

> this cultic world of sound the instrument [the panpipe] … causes a well-ordered tone series to be heard; and with the unity of syrinx and nymph Ovid described the goal towards which the tone series moves, which has always been a drawing of lines in the invisible.
>
> (1060)

Bloch accordingly interprets the pathos of Pan's lament for the absent Syrinx as the prefiguration of a revolutionary call to which the chiliastic Marxism espoused by Bloch responds. See my "The Wager of Imagination and the Logic of Hope," in *Paul Ricoeur in the Age of Hermeneutical Reason: Poetics, Praxis, and Critique*, ed. Roger W. H. Savage (Lanham, MD: Lexington Books, 2015).

18 Bloch, *Principle of Hope*, vol. 3, 1060.

19 Jean-Luc Nancy, *Listening*, trans. Charlotte Mandell (New York: Fordham University Press, 2007), 4.

20 Nancy, *Listening*, 20.

21 Steven F. Walker, "Mallarmé's Symbolist Eclogue: The 'Faune' as Pastoral," *PMLA* 93, no. 1 (1978). During Stéphane Mallarmé's day, "naked nymphs and satyrs were all the rage" (107). According to Walker, there is no obvious influence of Bouguereau's *Nymphes et Satyre* on Mallarmé's *L'après-midi d'un faune*.

22 See Walker, "Mallarmé's Symbolist Eclogue," 106.

23 See Paul Ricoeur, "Wonder, Eroticism, and Enigma," in *Sexuality and the Sacred: Sources for Theological Reflection*, eds. James B. Nelson and Sandra P. Longfellow (Louisville, KY: John Knox Press, 1994); Richard Kearney, "Ricoeur's Wager of the Flesh," in *Paul Ricoeur in the Age of Hermeneutical Reason: Poetics, Praxis and Critique*, ed. Roger W. H. Savage (Lanham, MD: Lexington Books, 2015).

24 Claude Debussy, "Monsieur Croche the Dilettante Hater," trans. B. N. Langdon Davies, in *Three Classics in the Aesthetics of Music* (New York: Dove Publications, [1928] 1962), 22–23. For Debussy, the word "ornament" had "nothing whatever to do with the meaning attached to it in the musical grammars" (23). Consequently, the musical arabesque—exemplified by Bach's violin concerto in G—was the basis for all forms of art. See William W. Austin, "The History of the Poem and the Music," in *Prelude to "The Afternoon of a Faun": An Authoritative Score Mallarmé's Poem, Backgrounds and Sources, Criticism and Analysis* (New York: W. W. Norton, 1970).

25 Debussy, "Monsieur Croche," 7; see Ronald L. Byrnside, "Musical Impressionism: The Early History of the Term," *The Musical Quarterly* 66, no. 4 (1980): 536; C. Henry Phillips, "The Symbolists and Debussy," *Music & Letters* 13, no. 3 (1932). Paul Valéry set out the idea of *poésie absolue*, according to which the poem is a self-signifying totality, to legitimate the principle of *l'art pour l'art*. See Carl Dahlhaus, *The Idea of Absolute Music,* trans. Roger Lustig (Chicago, IL: University of Chicago Press, 1989). Dahlhaus is right to suggest that the "dialectic at the root of both *poésie absolue* and absolute music" (55) highlights the fact that the work withdraws from the practical order of everyday existence into its own inner world. However, by equating the work's retreat from the real with its metaphysical elevation, Dahlhaus overlooks the work's mimetic character.

26 Jonathan D. Kramer, *The Time of Music: New Meanings, New Temporalities, New Listening Strategies* (New York: Schirmer Books, 1988), 1.

27 Kramer, *Time of Music*, 2.

28 Kramer, *Time of Music*, 46. In retrospect "we realize that an unchanging principle of organization" (42) supersedes the linear progression of tones and harmonies. Kramer additionally cites Schumann's *Stücken* from the *Album for the Young* (1848) and Chopin's *Prelude* in C Major, opus 28, no. 1 (1839) as examples of nonlinear tonal music.

29 Kramer, *Time of Music*, 46.

30 Ibid., 151.

31 Kramer, *Time of Music*, 151. According to Kramer, absolute time can be conceived as the ordered "linear succession of now-points." Claiming that absolute time "is measurable, whether by beats, pulses, or seconds" equates the rectilinear dimension of successive elements with the chronometric order of time.

32 Kramer, *Time of Music*, 3.

33 Ibid., 151.

34 Kramer, *Time of Music*, 152; see Robert S. Hatten, "A Theory of Musical Gesture and its Application to Beethoven and Schubert," in *Music and Gesture,* eds. Anthony Gritten and Elaine King (Burlington, VT: Ashgate, 2006). Hatten rightly defines gesture as the temporal shaping of a meaningful phenomenon. For him, a human gesture is "any energetic shaping through time that may be interpreted as significant" (1; original in italics). According to Hatten, these energetic shapings therefore can be interpreted indexically, iconically, and symbolically (4). See also David Lidov, "Emotive Gestures in Music and Its Contraries," in *Music and Gesture*; Ann Cox, "Hearing, Feeling, Grasping Gestures" in *Music and Gesture*, Cox rightly asks: "How is it that music makes us feel anything at all?" (45). However, the mimetic hypothesis that she adopts privileges the kind

of imitative engagement evinced by toe-tapping, for example, at the expense of a consideration of music's expression-exemplification of feelings and moods.

35 Kramer, *Time of Music*, 202.

36 Kramer, *Time of Music*, 45. Kramer adopts a left brain-right brain model to account for the differences between linear and nonlinear logics of the species of musical time.

37 Kramer, *Time of Music*, 55.

38 Kramer, *Time of Music*, 376. Hence for Kramer, a "vertically conceived piece ... does not exhibit large-scale closure" (55).

39 Kramer, *Time of Music*, 55.

40 Ibid., 378.

41 Kramer, *Time of Music*, 378; original in italics.

42 Kramer, *Time of Music*. The parallels between the different species of time and evolutionary *Umwelts* underscore the hierarchical ranking of Kramer's typology. Furthermore, just as each species of time has its characteristic features, each evolutionary level or temporal *Umwelt* has its laws and moods. The fifth species, "vertical time" is corollary with atemporality, which in Fraser's system is the most primitive of time's evolutionary levels. The corresponding mood of this *Umwelt*, in which simultaneity reigns, is that of a vast emptiness. Moment time (the fourth species) corresponds to the prototemporal *Umwelt* of statistical probabilities, where form "is understood cumulatively as the totality of constituent events" (395). The third species, multiply-directed time, has as its counterpart the *Umwelt* of eotemporality, where the relation of cause and effect is symmetrical so that for "any action-reaction pair, either event may be thought of as the action and either as the reaction." Hence, a "past-present-future ordering is indistinguishable from a future-present-past ordering" of relations. Nondirected linear time (the second species) has as its analogue the biotemporal *Umwelt* of beings that are born and die. Finally, the first species, directed linear time corresponds to the nootemporal "*Umwelt* of man," (396) in which a sense of self and personal identity joins memories and expectations to personal histories of beginning and endings.

Surprisingly, no species of time in Kramer's typology corresponds to the sixth level postulated by Fraser, namely sociotemporality. The sociotemporal *Umwelt* is that of "cultures and civilizations. Its science is sociology" (396). Each of the other *Umwelts* also has its science. Einstein's special theory of relativity is the science that investigates the *Umwelt* of atemporality. Quantum mechanism is the science of prototemporality. The sciences of the eotemporal *Umwelt* are that of the physical world of Newtonian mechanics and general relativity; the sciences of the biotemporal *Umwelt* are biology and physiology; the science of the nootemporal *Umwelt* is psychology.

43 Kramer, *Time of Music*, 394. The break between the hierarchical levels of moment time (prototemporality) and vertical time (atemporality) is therefore more radical than Kramer suggests. Although moment time approaches a sense of stasis in each of its moments, it relies upon a form of cumulative listening that constructs a mental image of proportioned relations. In vertical time, past, present, and future ostensibly dissolve into a timeless present.

44 Saint Bishop of Hippo Augustine, *Confessions*, trans. Richard Sydney Pine-Coffin (Middlesex: Penguin Books, 1961), 264.

45 Augustine, *Confessions*, 278; see Paul Ricoeur, *Time and Narrative*, vol. 1, trans. Kathleen McLaughlin and David Pellauer (Chicago, IL: University of Chicago Press, 1984), 22 ff.; Paul Ricoeur, *Time and Narrative*, vol. 3, trans. Kathleen Blamey and David Pellauer (Chicago: University of Chicago Press, 1988). Augustine

> has no other resource when it comes to the cosmological doctrines than to oppose to them the time of a mind that distends itself. This mind has to be that of an individual soul but by no means that of a world soul. Yet his meditation on the beginning of Creation leads Augustine to confess that time itself had a beginning along with created things.
>
> (244)

46 Ricoeur, *Time and Narrative*, vol. 1, 19; original emphasis: "But the impression is in the soul only inasmuch as the mind *acts*, that is, expects, attends, and remembers" (19).

47 Ricoeur, *Time and Narrative*, vol. 1, 19.

48 Ricoeur, *Time and Narrative*, vol. 1, 20. Ricoeur explains that the *distentio*, which Augustine opposes to the "present intention" to recite the psalm in its entirety, consists in this noncoincidence of attention (the present), memory (the present of the past), and expectation (the present of the future).

49 Ricoeur, *Time and Narrative*, vol. 1, 21.

50 Ibid.

51 Ibid.

> Is the *distentio* related in any way to the passivity of the impression? It would seem so, if this beautiful text, from which the *affectio* seems to have disappeared, is compared to the first analytical sketch of the act of reciting (27: 36). There the impression appears to be still conceived of as the passive reverse side of the very "tension" of the act, even when silent, of reciting: something remains (*manet*) insofar as we "can go over [peragimus] poems and verses and speech of any sort in our minds."... Thus, if we compare, as I believe we can, the passivity of the *affectio* to that of the *distentio animi*, we must say that the three temporal intentions are separate from one another to the extent that the intentional activity has as its counterpart the passivity engendered by this very activity and that, for lack of a better name, we designate as impression-image or sign-image. It is not only these three acts that do not coincide, but also the activity and passivity which oppose one another, to say nothing of the discordance between the two passivities, the one related to expectation, the other to memory. Therefore, the more the mind makes itself *intentio*, the more it suffers *distentio*.
>
> (20–21)

52 Ricoeur, *Time and Narrative*, vol. 1, 26. Ricoeur emphasizes that here "[t]hought is entrusted with the task of forming the idea of the absence of time in order to think time through as far as possible as that which passes" (25).

53 Ricoeur, *Time and Narrative*, vol. 1, 26. Ricoeur stresses that "[a]ntecedence must be thought of as superiority, as excellence, as supreme height." From this vantage-point, the

> eternal present declares itself in the first person: *sum*, not *esse*. Here again, speculation is inseparable from the recognition of the one who declares himself. It is in this that it is inseparable from the hymn. In this sense, we can speak of an experience of eternity in Augustine.... [I]t is precisely this experience of eternity that has the function of a limiting idea, when the intelligence "compares" time with eternity. It is the recoil effect of this "comparison" on the living experience of the *distentio animi* that make the thought of eternity the limiting idea.

54 Ricoeur, *Time and Narrative*, vol. 1, 24.

> What is posited, confessed, thought is in one stroke the contrast of eternity with time.... The anteriority of eternity with respect to time—in a sense of

> anteriority that remains to be determined—is given in the contrast between "something that exists that was not created" and something that has a before and an after that is subject to "change" and to "variation" (4: 6).
>
> (23)

55 Ricoeur, *Time and Narrative*, vol. 1, 7.
56 Augustine, quoted in Ricoeur, *Time and Narrative*, vol. 1, 7.
57 Ricoeur, *Time and Narrative*, vol. 1, 22.
58 Ricoeur, *Time and Narrative*, vol. 1, 26; see 22. Ricoeur emphasizes that the "absence of eternity is not simply a limit which is itself negative on the psychological level" (24).
59 Ricoeur, *Time and Narrative*, vol. 1, 22; see 28.
60 Paul Ricoeur, *Freedom and Nature: The Voluntary and the Involuntary*, trans. Erazim V. Kohák (Evanston, IL: Northwestern University Press, 1966), 453.
61 Ricoeur, *Time and Narrative*, vol. 1, 28.
62 Ricoeur, *Time and Narrative*, vol. 3, 15. For Aristotle, the perception of different instances rests on the "perception of the continuities of magnitude and movement, and on the relation of order between the before and after" (16).
63 Ricoeur, *Time and Narrative*, vol. 3, 19; see 18. Ricoeur stresses that Augustine's psychology does not refute Aristotle's cosmology.
64 Ricoeur, *Time and Narrative*, vol. 3, 40.
65 Paul Ricoeur, *Memory, History, Forgetting*, trans. Kathleen Blamey and David Pellauer (Chicago: University of Chicago Press, 2004), 114. Hence, "we can arrive at the absolute constitution of the flow only correlatively ... with the constitution of something that endures" (113). According to Ricoeur, two intentionalities therefore intertwine. The transverse intentionality "targets the thing that endures.... [O]ne then speaks of the retention of the tone" (113). The other intentionality aims at the self-constitution of the unitary flux. See Jacques Derrida, "The Voice That Keeps Silence," in *The Sound Studies Reader*, ed. Jonathan Sterne (New York: Routledge, 2012). Contra Husserl, for Derrida, "'Time' cannot be an 'absolute subjectivity' precisely because it cannot be conceived on the basis of a present and the self-presence of a present being" (501). The "'world" is therefore implied in the movement of temporalization, so that the temporalizing of temporality marks the limit of phenomenology's power to account for this movement. This limit thus marks the place that the enigma of time's inscrutability has with respect to a constituting consciousness (see 501–502).
66 Immanuel Kant, *Immanuel Kant's Critique of Pure Reason*, trans. Norman Kemp Smith (London: MacMillan and Co., 1929), 77; quoted in Ricoeur, *Time and Narrative*, vol. 3, 47.
67 Ricoeur, *Time and Narrative*, vol. 3, 50.
68 Ibid., 57; see 54.
69 Ricoeur, *Time and Narrative*, vol. 3.

> To connect the authentic structure of time to that of Care is, immediately, to remove the question of time from the theory of knowledge and to bring it to the level of a mode of being that (1) retains the scar of its relation to the question of being; (2) has cognitive, volitional, and emotional aspects, without itself being reduced to any one of these, or even being situated on a level where the distinction between these three aspects is pertinent; (3) recapitulates the major existentials as projection, thrownness into the world, and fallenness; and (4) provides a structural unity for these existentials that straight-away posits the requirement of "Being-a-whole" (*Ganzsein*) that leads directly to the question of temporality.
>
> (63–64)

70 Ricoeur, *Time and Narrative*, vol. 3, 70; see Martin Heidegger, *Being and Time*, trans. John Macquarrie and Edward Robinson (New York: Harper and Row, 1962); Martin Heidegger, *The Concept of Time*, trans. William McNeill (Oxford: Blackwell, 1992).
71 Ricoeur, *Time and Narrative*, vol. 3, 72.
72 Ricoeur, *Time and Narrative*, vol. 3, 65. Ricoeur accordingly asks

> whether the entire analysis of temporality is not tied to the personal conception that Heidegger has of authenticity, on a level where it competes with other existentiell conceptions, those of Pascal or Kierkegaard—or that of Sartre—to say nothing of that of Augustine. It is [sic] not, in fact, within an ethical configuration strongly marked by a certain Stoicism, that resoluteness in the face of death constitutes the supreme test of authenticity? More importantly, is it not within a categorical analysis, heavily influenced by the recoil-effect of the existentiell on the existential, that death is held to be our utmost possibility, even our ownmost potentiality, inherent in the essential structure of Care?
>
> (67)

73 Ricoeur, *Time and Narrative*, vol. 3, 91.
74 Ibid., 92.
75 Ibid., 92; see 81–82.
76 Ricoeur, *Time and Narrative*, vol. 3, 244. For Heidegger, the dividing line between these two perspectives on time guards against authentic temporality's leveling; for Ricoeur this fracture gives rise to an aporia that calls for a mimetic rejoinder.
77 Ricoeur, *Time and Narrative*, vol. 3, 250.
78 Ibid., 193.
79 Ricoeur's borrows the terms "space of experience" and "horizon of expectation" from Reinhart Koselleck's *Future's Past: The Semantics of Historical Time*, trans. Keith Tribe (Cambridge, MA: MIT Press, 1985).
80 Ricoeur, *Time and Narrative*, vol. 3, 259.
81 Ricoeur, *Time and Narrative*, vol. 3, 258; see my "Judgment, Imagination and the Search for Justice," *Études Ricoeuriennes/Ricoeur Studies* 6, no. 2 (2015).
82 Ricoeur, *Time and Narrative*, vol. 3, 261.
83 Ibid., 269.
84 Martin Heidegger, quoted in Ricoeur, *Time and Narrative*, vol. 3, 269.
85 Ricoeur, *Time and Narrative*, vol. 3, 241. Privileging narrative in this way highlights the connection between narrative identity and the narrative art's refiguration of time.
86 Ricoeur, *Time and Narrative*, vol. 3, 271.
87 Ibid., 263.
88 Ibid.
89 Ibid.
90 Ricoeur, *Time and Narrative*, vol. 3, 264: "The permanence of the *aion* [then] becomes that of an eternally immobile identity."
91 Anaximander, quoted in Ricoeur *Time and Narrative*, vol. 3, 263.
92 Ricoeur, *Time and Narrative*, vol. 3, 265.
93 Ibid., 266.
94 Ibid.

3 To the glory of God

Johann Sebastian Bach's declaration that music's sole purpose was for the glory of God alone (*Soli Deo Gloria*) tasks music with a sublime vocation. Through giving voice to the heart's desire, music inclines the soul toward its spiritual refreshment. Bach's conviction that fanciful arrangements of tones were only a "devilish blare and hubbub"[1] apart from this higher calling accentuates the gravity of music's affective power within Pietistic and Lutheran traditions. Luther, who praised music's spiritual value, advised that "next to the Word of God, the noble art of music is the greatest treasure in the world."[2] Music had no equal among the arts. For only music could quiet and cheer the soul by freeing it of troubling worries and thoughts. The prophets were similarly drawn to music more than any other art inasmuch as they proclaimed the truth through psalms and hymns of lament and praise.[3] When placed in the service of worship, music's power was second only to the Word's prophetic or kerygmatic proclamation. According to Luther, some of whose chorales Bach set in his cantatas and *Passions*, music therefore encouraged women and men to praise their Creator and to profess their soteriological hope.

The idea that Bach's sacred music has a distinctively religious character that sets it apart from his secular works immediately raises the question as to how musical expressions of the sacred and secular differ. On the one hand, attributing the difference between sacred and secular expressions to a work's intended function risks subordinating music's affective force to the contexts and purposes of the work's performance. Conversely, what reason would there be to include music in a religious service apart from its expressive power? We might wonder why, for Luther as for the prophets, songs of hymns and praise would have a place in religious worship if music did not speak to the heart. John Calvin similarly extolled the value of singing psalms to rouse the worshiper's heart and to incite it to render its praise and thanks.[4] Placing music in the service of a religious text might seem to be an obvious way of distinguishing between sacred works and secular instrumental compositions. But then we should also ask whether the reverse is not also true: Does music color the text's reception through the ways that its affective tonalities prepare the heart (θυμός) for its conversion?

The question as to whether devotional music inclines listeners toward receiving a religious text's spiritual or theological message is the staging ground for my reflections on the sense of eternity in Bach's music. This sense of eternity bears the trace of the Hebraic inspiration that opposes the transitory character of our mortal lives to God's eternal presence. The evocation in Bach's music of the presentiment of abiding grace resists reducing this sense of eternity either to the temporal permanence of an immutable present or to the sempiternal return of the same. In the last two chapters, I indicated how music's temporal features and affective attributes arise from the operation that draws expressive figures from successive heterogeneous elements. I then identified this operation with music's poetic response to the aporias that arise from the two ways of thinking about time (cosmological and phenomenological). In this chapter, I want in part to consider how eternity functions as a limit-condition with respect to time's ultimate unrepresentability. As the emblem of the nonbeing of time, eternity is the hallmark of time's inscrutable character. Expressions of a "time beyond time" that obliterate movement and change are more Greek than Hebraic in this regard. The aporias engendered by phenomenological time's and cosmological time's mutual occultation therefore serve as a caution against reducing eternity to a function of the ordinary concept of time. Accordingly, eternity marks out the horizon of the other that in religious experiences is the object of our ultimate concern.

Religious aesthetics and the feeling of height

By identifying the sense of eternity in Bach's music with his music's evocation of the presentiment of abiding grace, I am deliberately situating my reflections within a broader inquiry into music's mimetic efficacy. My thesis, that music revalues time through exploring ways in which time can be surpassed by its other, would be vacuous if it were not for the fact that music affects the ways that we inhere in the world through expressing feelings and moods that are unique to it. Music opens us to the world and the world to us by modulating our affective affinities with ways and modalities of feeling, thinking, and acting. Music's ontological vehemence—by which I mean its power to shatter congealed habits and practices by giving voice to sentiments and moods exemplified by musical works—takes root in a work's expressive power. In answer to the enigma of time's inscrutability, music revalues time by staking out the borderlines of the alterity of an order that is superior (in a sense that still needs to be clarified) to that of the being of time.

The sense of assurance that I will attribute to the presentiment of abiding grace in Bach's music has a decidedly secular advocate in Theodor W. Adorno. For Adorno, the "promise of objective salvation"[5] in Bach's music stems from the way that the polyphonic organization of independent voices supports the harmonic progressions. Bach's paradigmatic mastery of harmonic and contrapuntal impulses is the emblem of the emancipated subject's ultimate happiness, insofar as the security of the fit between harmonic and contrapuntal forces symbolizes the state of grace that the subject hopes to attain. On this interpretation, the E-flat major fugue's affirmative tone (from *The Well-Tempered Clavier* Book 2 [BWV 876]) presages the ideal of a comprehensively objective, absolute order of freedom beyond the world's fallen state. According to Adorno, Bach's music evokes this ideal not by mirroring the solitary subject as the "guarantee of meaning, but [by aiming at this solitary subject's] ... abolition and transcendence."[6]

The state of freedom aesthetically prefigured by Bach's music is thus the watchword of an emancipatory alternative that marks out the horizon of a demythologized world's hope for redemption. Ernst Bloch, whose chiliastic Marxism Adorno emphatically rejected, similarly regards Bach's music as an anticipatory expression of the world's salvation. The *sursum corda* ("lift up your hearts") that resounds in Bach's organ fugues summons the note of praise for the coming reign of the kingdom that brings an end to suffering on earth. The confidence in the victory of peace inspired by Bach's music, along with its unparalleled expressions of the fear of and longing for death, as well as hope and consolation, gives voice to a range of feelings that Bloch regards as belonging to the realm of religious emotions. However, for Bloch the mythical subject of Bach's sacred music—and especially the ultimate sacrificial act for humanity's redemption—is only an allegory of the human desire for deliverance from the evil of suffering. The notion that Bach's music is intellectual "sexton's music without a spiritual tone"[7] belies its expression of pain, weariness, and despair as well as its anticipation of the joy of the heavenly kingdom that Bloch believes Marxist science will bring about here on earth. The aesthetic fount of a secular soteriological promise, Bach's music presages the victory of the *nunc aeternum* over the violence of history and the vagaries of time.

The promise of salvation that these two proponents of liberatory social projects seize upon eschews any specifically religious overtones. Instead, the note of hope that Adorno and Bloch hear in Bach's music takes root in their respective critical social sciences' emancipatory ambitions. For Adorno as for Bloch, Bach's music offers a presentiment of humanity's redemption in a utopian order that would bring the history of a world that is punctuated by social injustices to an end. At the same time, the claim to be absolutely radical raised by each of these critical social theorists confounds their respective projects. Whereas Adorno's critique loses itself in a relentlessly negative dialectic, Bloch's eschatology abandons its historical handholds.[8] The impossibility of mounting a comprehensive critique that would account for the theorist's place in it lays bare the failure of a totalizing reflection. The hubris of critiques that claim to be absolutely radical consequently masks the practical task of preserving the tension between our expectations and the exigencies that in our experiences give rise to as yet unfulfilled demands. In turn, the practical character of this task bears out the fact that—far from being in a position of sovereign mastery—we act in situations that we did not choose the consequences of which we cannot always control or foresee.

The preceding remarks on Adorno's and Bloch's interpretations bear directly on my inquiry into the significance of the enigma of time's ultimate inscrutability with respect to the religious character of Bach's music. The critique of critiques that claim to be absolutely radical highlights the ethical and political stakes of a hermeneutical form of reasoning that refutes the subject's self-foundational claim.[9] Correlatively, the hermeneutical refutation of the subject's claim to posit itself brings to the fore the agent's relation to, and dependence upon, a history of which she is a part. The experience of belonging to this history is a condition for exercising the powers and capacities that she has and which are inscribed in the human condition. Far from immobilizing us, the feeling of dependence that accompanies this experience of belonging is a source of the convictions that we hold regarding our obligations to those who came before us, and to the heritages maintained through the succession of generations. From this vantage point, the experience of being affected by history and of belonging to it is the staging ground of the initiatives that we take (as I will explain in subsequent chapters).

The question that interposes itself at this point is whether the feeling of dependence that is rooted in the experience of being affected by the histories of which we are a part has a radical antecedent in the absence of aseity that I have previously suggested bears the stamp of the ontological difference between time and eternity. I wonder in fact whether Adorno's and Bloch's ostensible demystifications of the promise of salvation in Bach's music occlude the inner connection between the presentiment of an abiding grace and the sublimation of the deficiency that places its negative imprint on our experiences of time. The affinity of this presentiment of eternity with religious feelings is unmistakable. Yet if there is anything to be gainsaid by deconstructing critiques that claim to be absolutely radical, it is that religious feelings cannot provide an alternative foundation for claiming to know the absolute. On the contrary, in light of the absence of aseity evidenced by our inability to master time, the truth of religion springs from the avowal that our existence is in some way given to us. Only then is the "feeling of absolute dependence"[10] the ground from which all religious attitudes proceed.

This feeling of absolute dependence accompanies the recognition of an order of passivity that is more radical than our being affected by history. Paul Tillich points out that in its most basic sense, religion is the "dimension of depth"[11] that manifests the ground of our ultimate concern. This ultimate concern brings us face to face with the inscrutability of the unconditioned, or of the infinite. The vis-à-vis with the unconditioned is accordingly the ritual staging ground of basic feelings that are at once both ethical and cosmic.[12] The unique vernaculars of religious languages and rites clearly inform the ways that members of different religious communities understand their place in the world. Furthermore, fundamental feelings associated with religious practices and beliefs are always experienced within the context of a language that has its own customs and history.[13] Religious violence takes hold when different faith traditions each lays claim to the absolute. Marianne Moyaert identifies this violence with the effort to domesticate the vanishing point of religious experience, namely the illimitable condition of a foundation that is without foundation.[14] The abyss opened up by the disproportion between our finite capacities to give expression to the unconditioned and this illimitable condition thus paves the way for the ontotheological desire for human mastery (as I will also argue in Chapter 4 in connection with my critique of tonality's ontotheological resonances).

The feeling of height, that in the region of religious aesthetics takes hold in the feeling of the sublime, countermands the temptation to possess the absolute. The anteriority, exteriority, and superiority of the object that in religious experience is at the heart of one's ultimate concern reveal themselves through this feeling. Faith takes root in the way that the object of one's ultimate concern orients one's choices. Ricoeur therefore stresses that the feeling of absolute dependence "responds to an initiative that always precedes me"[15] in the manner of an invitation or injunction. The pre- or hyperlinguistic characteristic of an interpretation, the ceaseless movement of which "begins and ends in the risk of a response that is neither engendered nor exhausted by commentary,"[16] marks out a region where the experience of the sacred's "numinous" power can properly be said to be aesthetic. The sacred's inscription in an experience that has not yet been brought to the level of language (as in the case of hierophanies where the sacred manifests itself in rocks or trees venerated by believers) thus reserves a place for a feeling of height that belongs to a level of experience that precedes, as it were, the proclamation of a religious or theological message.

The difference between the manifestation of the sacred in hierophanies and the proclamation of doctrinal beliefs brings us face to face with the difficulty that Jaroslav Pelikan raises when he points out that the difference between Bach's sacred and secular music cannot be demonstrated empirically. How, in the absence of a religious text, could we account for the feeling of devotion evoked by a secular work such as Bach's *Air on the G String*? And would the addition of a text make the *Air* "more authentically 'devotional'"?[17] Conversely, is the aria "*Liebster Jesu, mein Verlangen* ['Dearest Jesus, My Desiring']" from the cantata BWV 322 no longer devotional when it is transcribed for instrumental performance? For Pelikan, the "pietistic or romantic subjectivism that professes to be able to discern Bach's Lutheran orthodoxy in his 'sacred' music"[18] is hardly credible. Hence, the question: Can the devotional character of Bach's music be credited to its affective qualities or is it dependent upon the content of a message spelled out by a text?

Rather than draw us back into the ideological maelstrom over music's contested relation to language, the question of the source of this devotional character opens a new front in my hermeneutical investigation into the relation between music's significance and time's ultimate unrepresentability with regard to the feeling of the sublime. In the Introduction, I pointed out how romantic sensibilities capitalized on the claim to aesthetic autonomy by vesting instrumental music with its metaphysical dignity. Kretzschmar launched his program for a musical hermeneutics on the grounds that music stands in need of an interpretive supplement in order that listeners understand the composer's intended meaning. In response to Hanslick's efforts to place aesthetics on a scientific footing, Kretzschmar set the stage for deploying interpretive strategies as a means of decoding the significance of formal structures and processes. However, exorcising the epistemological and even ideological precepts that set music against language as the means of exteriorizing an internal thought or image only reverses the metaphysical connection that in absolute music constituted the basis for the intimation of the absolute, even when the linguocentric predicament formulated by Charles Seeger had presumably been unseated. The sublime, and correlatively the ineffability of the unconditioned or the absolute, accordingly comes under a deconstructive hammer that itself is hostage to the effects of the subjectivization of expression.

By suggesting that the feeling of the sublime has a place in religious aesthetics, I am by no means insisting that expressions of the sublime carry the weight of religious convictions and injunctions. Kant in his *Critique of Judgment* makes it clear that the feeling of the sublime rests on a purely aesthetic judgment. For him, the sublime can only "be found in a formless object, insofar as we present *unboundedness* either [as] (sic) in the object or because the object prompts us to present it"[19] in its illimitable boundlessness. Kant emphasizes that the feeling of the sublime is initially contra-purposive with respect to the imagination's power to provide an adequate representation of the object that evokes this feeling in us. The feeling of respect elicited by an incomparable magnitude of something that is infinitely large or expansive (the mathematical sublime) or by the admiration aroused by nature's unbridled power (the dynamical sublime) accordingly springs from the way that the object of our liking awakens us to our supersensible vocation. Through attuning us to the thought of the unconditioned, the sublime not only harmonizes imagination with reason's ideas, but this mental attunement also conforms to the feeling produced by ideas for which the absolute whole is the determinate measure of reason's autonomous law. Hence, for Kant the feeling of respect is the moral sentiment that crowns the experience of the sublime. No object of the senses can therefore truly be called sublime, insofar as "reason demands absolute totality as a real idea."[20] The violence done to the imagination in striving to provide a sensible intuition of the object of our liking is the proof of the idea of our humanity, since sublimity can be found only in the mind. Hence, it is only by means of a certain subreption that respect for the object in nature replaces the idea of the humanity in us.[21]

Reason's demand for the unconditioned will be put to the test in Chapter 6, when I set in relief the war that Jean-François Lyotard maintains must be waged against the Enlightenment project. I will ask then how reputedly anti-narrative strategies in postmodern repetitive music elicit the postmodern sentiment of the sublime. Subsequently in Chapter 7, I will explore how aesthetic experience's lateral transposition onto the planes of ethics and politics guards against ethics' and politics' aestheticization. Only then will I be in a position to draw out the differences between moral and political injunctions and the force of the claims that accompany the aesthetic experiences occasioned by different works. The distinction that I am going to make between the feeling of height and religious aesthetics offers a glimpse of the critical requirement to give different regions of experience their proper due. As such, the connection, together with the differences between the feeling of height and religious aesthetics, offer an opportunity to explore the intersection between music's evocation of the unbounded in its expressions of a "time beyond time" and the feeling of absolute dependence in which religious attitudes are rooted.

The feeling of height that in Bach's music counterpoints the Pietistic experience of "sin as *Angst*" [22] bears a marked affinity with philosophies of testimony in this regard. For these philosophies, height is the name for the "glory of the infinite."[23] Glory is neither a phenomenon nor a theme. Rather, as Ricoeur explains, it is the "unsaid of the said"[24] that accompanies the assignment of responsibility. This assignment of responsibility comes from the other, in whose debt it places me. As such, the exteriority of the injunction that accompanies the assignment of responsibility stands as testament to the vertical relation between the self, others, and the Other. Exteriority and height therefore are conjoined. Where, for Ricoeur, Heidegger is unable to open a way from ontology to ethics, Levinas blocks the path leading in the opposite direction. Levinas's hyperbolic strategy of the unsaid—with its "an-archy without a beginning, [which is the sign of a] passivity of passivity"[25]—culminates in the philosophical excess that settles the fate of subjectivity at the expense of the self to whom responsibility is assigned. The ontological event attributed by Levinas to the irruption of the future clearly countermands the subject's self-same temporal unity. According to Levinas, this event makes the present a "rip in the infinite beginningless and endless fabric of existence."[26] The present as evanescence thus takes the place of the experience of a hypostatized time, which is inaccessible to any phenomenology. Contrariwise, only a self who is capable of maintaining herself (*ipseity*) can be summoned by the injunction coming from the Other, which results in our being enjoined. The word "glory"—as in Bach's dedication *Soli Deo Gloria*—thus designates the "epistemic status of [the] height"[27] of the Other as the object of praise.

The ontological event that Levinas identifies with the evanescence of the present resonates with the sense of eternity in which the Other's fidelity makes time the "event of our relationships with the Other."[28] In a similar vein, Ricoeur remarks that Augustine addresses the Eternal in the second person. In return, the "eternal declares itself in the first person: *sum*, not *esse*."[29] Far from abolishing time, this declaration intensifies the contrast between the glory of the infinite and the soul's peregrinations. The feeling of height in Bach's music also highlights the contrast between eternity and time, albeit without the weight of the moral injunction that accompanies the assignment of responsibility coming from the Other. Consequently, the presentiment of the glory of the infinite in Bach's music differs from the status assigned to the phenomenon of height in philosophies of testimony.

The presentiment of the glory of the infinite in Bach's music contrasts starkly with representations of eternity, the inspiration for which is more Hellenic than Hebraic. Karol Berger argues that the da capo form employed by Bach in his *St. Matthew Passion* acquires its theological significance as the emblem of eternity. Yet, similar to Kramer's derivation of vertical time from time's ordinary representation, Berger identifies the "cyclical time" of the da capo form (which he contrasts with the "arrow of time") with time's annihilation.[30] Berger attributes Bach's anachronism to the way that Bach supplants the past's, the present's, and the future's ordered succession with the "simultaneity of the present."[31] Accordingly, the seeming obliteration of the succession of past, present, and future occludes the slippage between them. We might legitimately ask whether Berger's formulation of "cyclical time" masks the aporia brought to light by Augustine's reflection on the *distentio animi* (see Chapter 2). Hence, we might wonder whether "Bach's evident desire to neutralize time … as in the C-Major Fugue, or to abolish it altogether, as in the St. Matthew passion's opening chorus,"[32] sets the representation of the eternal Now against time's regulated flow at the cost of eclipsing the enigma of time's inscrutability.

By crediting the feeling of height in Bach's music to the sublimation of time's deficiency, I therefore not only want to highlight the difference between religious injunctions and the affective power of Bach's music, but by identifying the presentiment of the glory of the infinite with the feeling of the sublime, I also want to accentuate how the assurance of abiding grace that crowns this presentiment is the emblem of eternity's victory over the Pietistic experience of sin, which Pelikan stresses is one of anxiety. Rooted in the knowledge of the certainty of one's death, this anxiety takes hold in the sting and grief of time's wearing effects. Neither the Romantic belief in the "redemptive power of tears"[33] (with which Pietism had a special affinity) nor the supposed evangelical *affectus* of mawkish literary references to the dying and dead body of Christ rises to the level of this existential vis-à-vis with the wound of our mortally finite condition. However deeply felt the experience of the illimitable by way of the feeling of the sublime might be, this experience is not a summons to responsibility. Rather music, like fiction and poetry, "intend[s] being, … not through the modality of givenness, but rather through the modality of possibility."[34]

The proximity in Bach's music between the feeling of height and religious feelings thus takes root in the sublimation of the grief born from the brevity of our mortal existence. His music's sublime quality is tantamount to an aesthetic affirmation of the contingency of an existence turned into a destiny when a life that is given to one is adopted as one's own. This affirmation prefigures the riposte to the conceit of thought to elevate itself to the level of the absolute.[35] Bach's music consequently constitutes the threshold of religious aesthetics inasmuch as the affirmation of the part that we have in an existence that is given to us, and the joy that we have in being, are both the source of a wisdom that laments the brevity of mortal life and the spring of praise that resounds in it.

Augustine's dilemma

Music's power to affect its listeners clearly poses a dilemma when it comes to channeling this power for devotional purposes. In his *Confessions*, Augustine acknowledges that setting music to words gives the words life: "If I am not to turn a deaf ear to music, … I must allow it a position of some honor in my heart."[36] At the same time, he finds it "difficult to assign … [music] to its proper place."[37] Giving music too much prominence runs the risk of paralyzing the mind by gratifying the senses. Conversely, approving the "custom of singing in church, in order that by indulging the ears weaker spirits may be inspired with feelings of devotion"[38] trades on music's capacity to affect its listeners in order to prepare the heart for its conversion. In order to guard against music's idolatrous pleasures, Augustine places his trust in the power of sacred words. When "I find the singing itself more moving than the truth which it conveys, … this is a grievous sin."[39] Then "I have become a problem to myself."[40] Yet "I realize that when they are sung these sacred words stir my mind to greater religious fervour and kindle in me a more ardent flame of piety than they would if they were not sung."[41] For Augustine, music thus not only leads the mind astray but also inspires feelings in the believer that direct the heart toward the object of its devotion.

Augustine resolves the dilemma with which he is struggling by confronting the "danger that lies in gratifying the senses and the benefits which … can accrue from singing."[42] While he "still finds some enjoyment in the music of hymns, which are alive with your [God's] praises, when I hear them sung by well-trained, melodious voices,"[43] he espouses the doctrine that music fulfills its sacred purpose only when placed in the service of proclaiming the kerygmatic word:

> when I remember the tears that I shed on hearing the songs of the church in the early days, soon after I had recovered my faith, and when I realize that nowadays it is not the singing that moves me but the meaning of the words when they are sung in a clear voice to the most appropriate tune, I again acknowledge the great value of this practice.[44]

Music consequently fulfills its sacred function through inclining the heart to rejoice in singing hymns and psalms of praise.[45]

Augustine's response to his susceptibility to music's sensuous gratification, however, only redoubles the question I raised earlier regarding the devotional character of Bach's music. Augustine remedies the problem by placing music's power to move him in the service of a religious text. Yet, it is the music that in its various modes enlivens hymns and songs of praise by stimulating the emotions by means of "some mysterious relationship"[46] between these modes and the feelings elicited by them. I indicated earlier that this relationship rests on the connection between music's temporal features and affective properties. Consequently, the feelings evoked spring from the operation that draws a figure from a sequence of tones based on the appropriate mode. Feelings thus differ from emotions in an important respect. Where feeling reveals the "*élans* of our being and … its pre-and hyper-objective connections"[47] with the world, things, and beings in it, emotion is a corporeal disturbance that relates to a value judgment regarding the pain that is to be avoided or the pleasure that is to be pursued. Emotions set our desires in motion; feelings register the affective nuances of things.[48] By designating "qualities felt *on* things, *on* persons, *on the world*,"[49] feelings reveal the ways that we are affected inwardly at the same time that they restore our complicity with the world. Rather than positing any "being," feeling fulfills its function by uniting an "intention toward the world and an affection of the self."[50] Feelings therefore can be expressed and communicated in cultural languages and aesthetic forms thanks to the way that this intentional aim and its affective correlate coincide.

Hymns of praise have a unique significance in this respect. By emphasizing how singing intensifies the power of words to move him, Augustine singles out feeling's role in modulating the *élans* of our being. For, through raising everyday feelings above themselves, music augments the range of our hyper-objective connections with the world and the things in it at the same time that it intensifies the ways that we are affected inwardly by the affective qualities of the objects on which these feelings are felt. Acts of praise attest to what is most worthy and laudable. Hence, praise directs itself toward the object of its devotion. In praising, Gadamer tells us, the "standard by which we evaluate and comprehend our existence is made manifest,"[51] so that in praising we rejoice in the object of our ultimate concern. The discourse of *agape*, Ricoeur therefore reminds us, is also one of praise. By daring to speak of the imperative's poetic usage, he highlights how the optative mode of hymns and benedictions draws near the height of *agape* that is celebrated in the beatitudes. The "commandment that … the lover addresses to the beloved: Love me!"[52] commends itself through the tenderness of the lover's entreaty. The strong evaluation that in praise is "uttered in the mode of song"[53] thus reveals the tie between the height of *agape* and the object that reigns over all other concerns.

The truth that according to Luther the prophets proclaimed through psalms and hymns of lament and praise thus has both a theological and a hyper-theological significance. Theologically, this truth is rooted in the kerygmatic value of these psalms and hymns' sacred texts. At the same time, like lamentations that speak of suffering and abandonment, hymns and songs of praise also attest to the frailty and fragility of human life. Lament and praise here transcend the specifically theological meaning that they have within the various religious traditions within which they figure. For laments and hymns of praise speak to the sorrow and joy of all human beings who are capable of affirming a life that is far from aseitous. Both the accusation of having been abandoned to a miserable fate and songs of praise sound out the range of tonalities through which the feeling of absolute dependence reveals itself. As such, laments and praise give voice to the depths and heights of feelings that can also assume a specifically religious significance.

Joy and sorrow have a unique value in this respect. In joy and sorrow, the good that we desire and the evil that we flee become fused with the vital forces that animate the *élans* of our being. Joy and sorrow are closely related to experiences of pleasure and pain. Yet these emotions, if we can still call them that, reveal modalities of our being sanctioned by the good to which we aspire in joy or the evil that overwhelms us in sorrow. Joy attests that we are capable of affirming the part we have in being despite the absence of aseity that places its stamp on our mortal existence. Sorrow and anguish are the underside of this spiritual joy (as I indicated in reference to Pietism's preoccupation with sin and death). In joy, the height of the feeling of belonging to being that Ricoeur maintains "ought to be the feeling that is most detached from our vital depth—what is absolute—in the strongest sense of the word—thus becomes the heart of our heart."[54]

The heart in question constitutes the fragile moment *par excellence* of the intermediary condition of our manner of inhering in the world. Where pleasure places its stamp of perfection on our bodily experiences, spiritual feelings reveal the affective character of our openness to an existence that is given to us. Happiness transcends the finite perfection of partial pleasures in aiming at the "most excellent form of pleasure itself."[55] As signs of happiness, spiritual feelings are a promise of an as yet unfulfilled order realized in the love (φιλία) of others and the devotion to exemplary ideals. The heart (Θυμός) is thus the "living transition from βίος (life) to λόγος [logos]"[56] that unites and separates sensuous desires from reason, "whose specific desire is ἔρως [Eros]."[57] Ricoeur stresses that the "Eros that goes toward being is also that which recalls being as its Origin."[58] Hence, for him, reason "is my 'decree' and my 'destination'—my *Bestimmung*—the intention according to which I can 'continue my existence.'"[59] This destination (*Bestimmung*) is animated by the mood (*Stimmung*) that orients it affectively. Accordingly, reason engenders "feeling as an openness to happiness"[60] in accordance with reason's demand for the unconditioned. The feeling of height in Bach's music counterpoints this demand through giving voice to the glory of the infinite. This presentiment of spiritual joy is thus the affective index of a superabundance of meaning, the source of which for Augustine and Luther as theologians is the origin of goodness and truth.[61]

St. Matthew Passion

The dramatization in Bach's *St. Matthew Passion* of the events leading to Christ's crucifixion according to the Gospel of Matthew (Chapters 26 and 27), which celebrates the culmination of the Lenten season, is a testament to the reach of his music's spiritual power. Along with Biblical passages, the text by Christian Friedrich Henrici includes Lutheran chorales and free verse. In contrast to Hermann Samuel Reimarus's rationalist theology, the *Passion* emphasizes the "voluntary character of the death of Christ as ultimately the result of his love rather than of human malice."[62] In Pelikan's view, Bach saw Christ's intention to suffer and die as the subject of the work's entire narrative. The mutual exchange that takes the place of the contrast between the "innocent Lamb of God … [and] guilty humanity"[63] symbolizes the promise of God's redeeming grace. From Pelikan's theological standpoint, the *Passion*'s portrayal of Christ's exemplary act demands that "we respond with an existential decision for or against the claims of the story to represent the nature of Ultimate Reality."[64] The church is therefore the "natural habitat of Bach's *Passions*."[65]

The *St. Matthew Passion*'s religious ethos and kerygmatic significance clearly fit the liturgical intent of the church's penitential commemoration of Christ's suffering and death in anticipation of the Easter event. This religious ethos is attributable in part to the narrative texture of the passion story. At the same time, the music's spiritual power bears out the conclusion to be drawn from the preceding discussion as to whether to credit the devotional character of Bach's music to its affective qualities, or whether to credit this devotional character to the content of an accompanying text. By clothing the *St. Matthew Passion*'s meditation on Christ's sacrifice in human dimensions, the opening chorus magnifies the pathos and the glory of the act of atonement. Here, in the call from the Daughter of Zion to the daughters of Jerusalem to weep for themselves and their children, the "Anselmic doctrine of redemption as satisfaction rendered through the blood of Christ"[66] (which for Pelikan runs through Bach's *St. Matthew Passion* as a "crimson thread"[67]) acquires its somber grandeur. This opening lament joins the "deepest contemplation of sin"[68] to a reflection on the magnitude of the cost of humanity's redemption. The gathering darkness of the *Passion*'s religious dramaturgy in anticipation of the Good Friday event is equally an incantation of the soul. A "ritual of the soul,"[69] "Kommt, ihr Töchter, helft mir klagen!" ("Come, ye daughters, help me lament!") dissolves the distance separating a guilty humanity from its soteriological hope (Example 3.1).

By setting the question of the devotional value of Bach's music in the context of a reflection on the *St. Matthew Passion*'s theological *affectus*, I also want to draw out how the presentiment of abiding grace that I have identified with the feeling of height evinces a sense of the eternal that is consonant with Bach's Lutheran heritage. Eric Chafe argues that Bach's *St. Matthew Passion* is "one of the most intricate and profound musico-allegorical

Example 3.1 J. S. Bach, "Kommt, ihr Töchter," *St. Matthew Passion*, mm. 1–5.

structures in all of music."[70] According to him, *catabasis* ("descent through the circles of fifths, [or] modulation in the direction of increasing flats"[71]), *anabasis* ("ascent; modulation toward increasing sharps"[72]) and, more broadly, the tonal architectures of Bach's music are allegorical Lutheran representations of Christian dogmatics and a life of faith. By identifying tonal structures and figures drawn from the doctrine of affects with the theological message that Bach purportedly intended to convey, Chafe proposes to extend traditional musical hermeneutics' interpretive reach by expressly thematizing the relation between the *Passion*'s affective power and its religious significance.

By insisting that music can reveal the ground of an eternal grace directly without reference to an associated text, Jeremy Begbie similarly reserves a place for the expression of religious feelings in Bach's music. For him, music can thus serve as an icon of the "glory of God."[73] Nevertheless, theological claims regarding music in the end must take refuge in concepts about which there is some agreement. Accordingly for him, the "trinitarian eternity"[74] evidenced by the *St. Matthew Passion*'s interlacing of a modern concept of history's inexorable advance and a sense of constancy (which John Butt attributes to the recurrences of repeating components) bears the stamp of a creationist theology.[75] From this standpoint, the contrast between time and eternity acquires its theological significance by reason of the way that the *St. Matthew Passion* emulates the Christological vision of a world drawn toward its fulfillment through the irruption of eternity within human history. While Begbie critiques the "myth of [time's] passage,"[76] his claim that arresting or restraining the directed movement of tension and release evokes a sense of eternity, however, redoubles the difficulties that spring from identifying music's temporal attributes with the ordinary concept of time.[77] The soteriological framework adopted by Begbie clearly brings the ordinary concept of time and eternity into conflict. Yet, as I have argued, this conflict acquires its existential—and, I would add, theological—import only by reason of the inability of phenomenology or cosmology to constitute time. The presentiment of abiding grace that crowns the sublimation of the anxiety wrought, in this religious context, from the consciousness of sin and death thus evinces the glory of the infinite as the origin of the good that we desire to be.

The text of the soprano aria, "Aus Liebe will mein Heiland sterben" ("For love my Savior now is dying"; *St. Matthew Passion* 58) encapsulates the *Passion*'s intended theme: "the Savior Jesus Christ suffered and died because of his love for humanity … to make it possible for the mercy of God to forgive sin and guilt."[78] Pelikan comments that Bach "seems to have been quite content … to work within the schema first systematized by Anselm"[79] of Canterbury (ca. 1033–1109). Hence the *Passion*'s prose and poetry, which is "set to music of almost unbearable emotional power,"[80] has as its theological counterpart not the "rational-apologetic treatise *Why God Became Man*, but an emotional-devotional treatise *Meditation on Human Redemption*."[81]

The emphasis on repentance and the moral earnestness of the believer's inner life in the *St. Matthew Passion* are indicative of Pietism's preoccupation with death. Ensconced within the liturgical concern with time and transiency as "sacralized by the four seasons of the Christian year,"[82] this preoccupation with death (for which Bach's funeral cantata "Gottes Zeit" [BWV 106] is a poignant reminder) is a testament to the *St. Matthew Passion*'s theological force. Hagiographies that portray Bach as a staunchly orthodox Lutheran often identify the spiritual qualities of his music with its representations of the Christian message. His music's soteriological import is accordingly often attributed to the kerygmatic content of the narratives and texts in the service of which it has been placed. Karl Barth pointedly singles out Bach's *St. Matthew Passion* for not proclaiming the Easter message. For Barth, the *Passion* fails to instill a sense of Christian hope, since "Jesus never once speaks in it as the Victor."[83] Pelikan by contrast maintains that the *Passion* is a "magnificent song of mourning for the fallen Savior and of gratitude of his life and death."[84] This song of mourning thus sets the *St. Matthew Passion* apart from the *Saint John Passion*, in which death and hell are vanquished.

The religious language in which the consciousness of sin and death is couched bears on the way that this soteriological hope is interpreted and expressed. The yearning that Pietism developed into a high art despite the difficulty of finding biblical warrant for its devotion to, and hunger for, "release from the prison of this life"[85] is a testament to the awareness of the impermanence of mortal life. As such, the Pietistic yearning for release has its cathartic counterpart in the penitent's identification with Christ's suffering and death. At the same time, the music's expressive power vests the *Passion*'s theological *affectus* with a vital force. Hermeneutically speaking, the joy and the sorrow to which the *St. Matthew Passion* gives voice are therefore sentiments *felt* on the good to which we aspire and the evil that we flee.

One of the most beloved of German Protestantism Lenten chorales, "O Haupt voll Blut und Wunden" figures prominently in the *St. Matthew Passion*. The earliest printed version of the chorale melody dates to 1601 in Nürnberg, where it appeared as the tune for the love song "Mein G'müth ist mir verwirret" ("My Mind is so Confused"). Gerhardt German adapted the original Latin text of the Lenten hymn "Salve caput cruentatum," which was once attributed to Bernard of Clairvaux and has been more recently ascribed to Cistercian Arnulf of Louvain.[86] Bach used the chorale melody in his cantata "Komm, du süsse Todesstunde" ("Come, O Death, Thou Sweetest Hour"; BWV 161). In the *St. Matthew Passion*, "O Haupt voll Blut und Wunden" appears five times: twice in connection with the scene on the Mount of Olives, once on the occasion of the trial before Pilate, once in connection with Jesus's scourging by the soldiers, and once following the moment of his death.

The chorale's recurrence at decisive moments in the narrative punctuates the *Passion*'s dramaturgical structure with its various settings' affective resonances. The chorale is heard first following the institution of the sacrament of the Last Supper. The chorus responds with a hymn of praise using the lines of the fifth stanza.[87] The sixth stanza follows soon after, following Peter's declaration that he would never deny Christ (in response to Christ's prophecy that Peter would betray him three times before the cock crowed).[88] The chorale appears a third time in connection with Christ's encounter with Pontius Pilate. In response to the Evangelist's words, "Und er antwortete ihm nicht auf ein Wort" ("And he gave him no answer, not even to a single charge"), the chorus commends the believer to remain steadfast in the confidence of God's abiding love using the first stanza of Gerhard's "Befiehl du deine Wege."[89]

The chorale's fourth appearance marks the nadir of Christ's humiliation. Following the Evangelist's words describing Christ's treatment at the hands of the soldiers ("Und speieten ihn an, und nahmen das Rohr, und schlugen damit sein Haupt." ["And they spit upon Him, and took the reed, and smote Him on the head."]), the chorus sings two successive stanzas.[90] This penultimate instantiation of the chorale has been recognized almost universally as both the most musically intriguing and theologically complex. The key scheme of the chorale's different settings has led to drawing analogies between the melody's harmonizations and the chorale's theological meanings, as Chafe's reading attests. The tonal schema through which the different settings of the chorale's various appearances progress (E major through E♭ major, D major, F major, to A minor cadencing on the dominant) thus has been related allegorically to the *Passion*'s dramaturgical structure. Pelikan, for example, remarks that this fourth appearance of the chorale "stands in the majestic and heartbreaking severity of only one flat."[91] The moods that he attributes to the different keys in which the chorale is set clearly stem from these different setting's affective tonalities. As such, the tonalities of these different settings suit the meaning of the text, as we might expect in light of Augustine's remarks on the way that putting words to music using the appropriate modes kindles a more ardent flame in the heart of the believer.

David Hill goes even further in his analysis of the fourth setting. His Schenkerian analysis brings to the fore the network of relations that I said before are constituted by the synthetic operation that draws expressive figures from sequences of sounding elements. Analysis, I emphasized then, is a second-order process predicated on this operation's schematization of music's temporal characteristics and affective attributes. Hill accordingly identifies structural features of the chorale's harmonization with the text's theological import by focusing on the way that the threat to this setting's overall tonal coherence underscores how the "tension between the appearance of Jesus's bleeding face and the honor that the singers pay to it is reflected in the music."[92]

In Hill's view, the contrast between the first and second stanzas' poetic allusions lays bare the penitent's belief in Christ's blamelessness. For him, the exchange between D minor and F major foregrounds the dramatic opposition between the scorn visited on Christ in the first stanza and the second stanza's hymn of praise. The intimation of the D minor opening underscores the profundity of this moment in the Passion story. When, with the commencement of the second strophe we reinterpret the chorale's overall setting as in F major, the exchange between D minor and F major throws the chorale's essential structure into confusion. This structural confusion—which for Hill is indicative of the narrative's crisis—sets the chorale's setting's theological profundity in relief. For where the first four phrases appear to support an analysis in D minor, the seventh and eight phrases cast the D minor reading into doubt by imposing an F major reading on the analysis, thereby motivating a repetition that "allows a re-hearing of the tune in the context of F major, as opposed to the initially anticipated D minor"[93] setting. This analysis underscores how this setting's affective power springs from the schematizing operation on which the significance of the exchange that Hill identifies with this setting's theological gravitas depends. Bach's harmonization is both structurally intriguing and uncannily expressive in this regard.[94] However much analysis provides a point of access to structural ambiguities and lacuna, the figuration of the sense of anguish and humiliation that Hill identifies with the chorale's D minor opening stands as a testament both to Bach's creative power and to the chorale's affirmation of the sense of abiding grace at the heart of the *St. Matthew Passion*'s theological message (Example 3.2).

The chorale's final appearance follows the Evangelist's words: "Aber Jesus schriee abermal laut, und verschied" ("Jesus, when He had cried again with a loud voice yielded up the ghost"[95]). Initially, this setting of the chorale seems to support Barth's criticism that the *St. Matthew Passion* fails to proclaim the Easter message. Yet, the prayerful plea that resounds in Bach's setting of the chorale is a cipher for the welcome by a penitent heart of the assurance of abiding grace. This assurance is the spring of the hope of the everlasting gift of love. For this hope, which is couched in the language of Lutheran and Pietistic theology, reaches toward the height of *agape* from which the command to love proceeds. Can we not then hear a note of praise in this penitent prayer, in which fragile human beings seek comfort in the thought that they will not be abandoned? Pelikan suggests that the D-major setting of the "Gloria in excelsis" from the *Mass in B Minor*, which represents the next key in the modulatory progression from this last setting of "O Haupt" in the *St. Matthew Passion*, befits the celebration of Lenten's season's triumphant conclusion. This projected modulation might even be one way of proceeding from "Nunc Dimittis to Gloria in excelsis."[96] Whatever significance we might attach to speculations regarding this projected modulation, similar to the benediction "Dona Nobis Pacem" that concludes Bach's *Mass in B Minor*, this last setting of the chorale commemorates the Lenten season's final moment in anticipation of the glory of the infinite that crowns *agape*'s illimitable height.

Example 3.2 J. S. Bach, "O Haupt voll Blut und Wunden," *St. Matthew Passion*, 63.

The glory of the infinite

The *St. Matthew Passion*'s celebration of the Lenten event clearly uses "O Haupt voll Blut und Wunden" to theological advantage. The secular origin of the chorale tune notwithstanding, the chorale settings fulfill the devotional role that Luther and Calvin assigned congregational singing through personalizing the impact of the *Passion*'s message. In view of the tune's origin as a love song, we might wonder whether the settings of "O Haupt" in the *Passion* are more authentically devotional due to these settings' poetic and religious texts. Conversely, we could ask whether the spirit of praise in the "Osanna in excelsis" from the *Mass in B Minor* is also evident in "Es lebe der König!," which Bach composed based on the same melody.[97] If Bach's music's devotional character were strictly a function of a text, what could we say about the presentiment that attests to the glory of the infinite by way of the feeling of height?

By inverting the relation between music and text, this question foregrounds music's mimetic efficacy. If, as I have argued, music prepares the heart for its conversion through evoking feelings that we can make our own, is the use to which music is put not a testament to its affective force? However disconcerting music's sensuous character was for Augustine, he embraced the supporting role into which music was cast when words are sung to the appropriate tune. We could ask whether the "weeping expressivity"[98] that Wilfrid Mellers attributes to Bach's chromatic "C Minor Prelude" (*Well-Tempered Clavier*, Book II) universalizes the "tears of all things" (*lacrimae rerum*) by rendering them incarnate in sound *sub specie aeternitatis*. Placing the tragic dimension of life's misfortunes under the sign of eternity as evidenced in Bach's music marks out the region of the aesthetic in which Bach's music converges with religious feelings. Furthermore, as I have already indicated in my discussion of the deficiency brought to the fore by placing our temporal existence against the backdrop of eternity, the feeling of absolute dependence in which these religious feelings take root bears the stamp of the absence of aseity from which this tragic dimension proceeds. In response, the sublimation of sorrow and anguish opens the way to spiritual joy. Bach's *St. Matthew Passion* gives voice to the glory of the infinite by way of the wonder of the feeling of height. As such, the presentiment of abiding grace that rules over Bach's music as the emblem of eternity stands at the threshold of religious aesthetics.

The presentiment of the assurance that we have a part in being despite the deficiency that strikes at the heart of time consequently evinces a sense of eternity that is more Hebraic than Hellenic. Like wonder, which "expresses the irruption of the 'other' into consciousness,"[99] the feeling of height in Bach's music offers the assurance of joy in the midst of life's trials and tribulations. To be sure, this assurance operates without injunction or constraint. The irruption of joy may not rise to the level of the ethical that Levinas credits to time's relation to the Other. Yet Bach's music gains its exemplary value in return through its evocation of joy's crowning note of grace.

In its own way, the "Andante" from Bach's *Sonata for Flute and Continuo* in E minor (BWV 1034; Example 3.3) gives voice to joy's embrace of the illimitable that in the realm of religious aesthetics is the spring of the feeling of sublime height. While it lacks the *St. Matthew Passion*'s religious solemnity and theological gravitas, the graceful air of the "Andante" clothes itself with this feeling. Far from expressing a cultured sentimentality, this insouciant air places its presentiment of abiding grace under the emblem of the glory of the infinite. The rising turn of phrase that brings the "Andante" to a close crowns this movement's evocation of a feeling of unburdened joy. As such, this crowning moment transfigures the pathos born from the passivity of an existence that is given to us into an openness to happiness.

Example 3.3 J. S. Bach, "Andante," *Sonata for Flute and Continuo* (BWV 1034), mm. 7–13.

That the ethos of this "Andante" has no determinate object on which the mood emanating from it can be felt ultimately brings to the fore the enigma of assigning a devotional character to Bach's music. On the one hand, his sacred works specify the object of devotion by setting a religious text to music. On the other hand, attributing the devotional character of this music to a religious or poetic text overlooks the power of his music to incline listeners' wills toward the object of their ultimate concern. Augustine's, Luther's, and Cavlin's concern with music's place in the life of a congregational community attests to the dilemma that the object of the feelings evoked by music are often indeterminate. For atmospheric feelings (*Stimmungen*) are in the end properties of individual works. As such, the devotional quality that a work possesses springs from the bearing that the feelings exemplified by the work have on our manner of inhering in the world. Such feelings are not governed by Ideas or by a sense of belonging to a greater order of which we are a part. Yet they constitute the ground of a range of feelings that are the schemata of our openness to an existence that we adopt as our own. This openness to happiness springs from a feeling of height that in Bach's music stands as an affirmation of joy beyond the sadness of our finite condition. The assurance that secular critical theorists and theologians alike hear in Bach's music is a testament to its enduring power. In Bach's music—secular and sacred—the glory of the infinite rules over the pathos of our mortal condition as the presentiment of abiding grace.

The sense of eternity that I have attributed to the glory of the infinite by way of this presentiment's evocation in Bach's music sets in relief the destruction of traditional tonal music's ontotheological resonances. In the next chapter, I ask whether Schoenberg's renunciation of tonal strictures prefigures the "bad infinity" of an instant of time frozen between two fleeing horizons. In contrast to the assurance of the part that we have in being, which in Bach's music resounds in response to the deficiency that places its stamp on time, Schoenberg's riposte to the Hegelian temptation to master time gives voice to the crisis of a present torn between the two fleeing horizons that bound the space between them. This temporally bounded space constitutes the site in which the initiatives we take distinguish the force of the present from both the past and the future. It remains to be seen whether the war waged on totality by postmodern deconstructions recasts the destruction of tonality's ontotheological value. In this respect, the difference between representations of the end of time and the mythic return to a time *in illo tempore* will prove to be a critical touchstone (see Chapter 6). The preceding reflections on the sense of eternity that in Bach's music reverberates with the presentiment of abiding grace consequently initiates a series of inquiries that lead to a broader consideration of music's significance vis-à-vis ethics and politics.

Notes

1 Johann Sebastian Bach, quoted in Christoph Wolff, *Johann Sebastian Bach: The Learned Musician* (New York: W. W. Norton, 2000), 309; see Hans. T. David and Arthur Mendel, eds., *The Bach Reader: A life of Johann Sebastian Bach in Letters and Documents*, rev. ed. (New York: W. W. Norton, 1966), 32–33.

2 Martin Luther, quoted in Walter E. Buszin, "Luther on Music," *Musical Quarterly* 32, no. 1 (1946): 83.

3 See Martin Luther, "Letter to Senfl 4 October 1530"; quoted in Buszin, "Luther on Music," 84:

> I am not ashamed to confess publicly that next to theology there is no art which is the equal of music, for she alone, after theology, can do what otherwise only theology can accomplish, namely, quiet and cheer up the soul of man, which is clear evidence that the devil, the originator of depressing worries and troubled thoughts, flees from the voice of music just as he flees from the words of theology. For this very reason the prophets cultivated no art so much as music in that they attached their theology not to geometry, nor to arithmetic, nor to astronomy, but to music, speaking the truth through psalms and hymns.

4 John Calvin, "Articles Concerning the Organization of the Church and of Worship at Geneva 1537," in the Rev. John Kelman Sunderland Reid, ed., *Calvin: Theological Treatises* (Philadelphia, PA: Westminster Press, [1954] 2000), 48; quoted in Jeremy S. Begbie, *Music, Modernity and God: Essays in Listening*, (Oxford: Oxford University Press, 2013), 12. For Calvin:

> [i]t is a thing very expedient for the edification of the Church, to sing some psalms in the form of public devotions by which one may pray to God, or to sing his praise so that the hearts of all be roused and incited to make like prayers and render like praises and thanks to God with one accord.
>
> (quoted in Begbie, *Music, Modernity and God*, 12)

5 Theodor W. Adorno, *Prisms*, trans. Samuel and Shierry Weber (Cambridge, MA: MIT Press, 1967), 138.

6 Adorno, *Prisms,* 138. According to Adorno, the

> duality of mind, harmonic and contrapuntal, circumscribing every one of the compositional problems that Bach paradigmatically solved, ... must exclude the image of him as the consummation of the Middle Ages. Were the image valid, he would neither have had the duality of mind, nor have struggled, especially in the speculative late works, with a paradox which would have been unthinkable for the old polyphonic mind, namely, how, in terms of *basso continuo* harmony, music could justify its progression as meaningful and at the same time organize itself polyphonically through the simultaneity of independent voices.... [I]n its [the E-flat major fugue's] substance—certainly not in its subjective consciousness—it is reflection on the happiness of musical security, the like of which is possessed only by the emancipated subject, for only it can conceive music as the emphatic promise of objective salvation.

See Theodor W. Adorno and Max Horkheimer, *Dialectic of Enlightenment*, trans. John Cumming (London: Verso, 1979).

7 Ernst Bloch, *Principle of Hope*, vol. 3, trans. Neville Plaice et al. (Oxford: Basil Blackwell, 1986), 1065.

8 See my "The Wager of Imagination and the Logic of Hope," in *Paul Ricoeur in the Age of Hermeneutical Reason: Poetics, Praxis, and Critique*, ed. Roger W. H. Savage (Lanham, MD: Lexington Books, 2015).

9 See my "Introduction: Paul Ricoeur and the Age of Hermeneutical Reason," in *Paul Ricoeur in the Age of Hermeneutical Reason: Poetics, Praxis, and Critique*, ed. Roger W. H. Savage (Lanham, MD: Lexington Books, 2015).

10 Paul Ricoeur, *Figuring the Sacred: Religion, Narrative, and Imagination*, trans. David Pellauer, ed. Mark I. Wallace (Minneapolis, MN: Fortress Press, 1995), 65; see Paul Ricoeur, "Lamentation as Prayer," in André LaCocque and Paul Ricoeur, *Thinking Biblically: Exegetical and Hermeneutical Studies*, trans. David Pellauer (Chicago, IL: University of Chicago Press, 1998). Ricoeur points out that

> [w]e may ... speak of religious experience to characterize attitudes toward the divine as different as a feeling of absolute dependence, the experience of a boundless trust, the thrust of ultimate concern, and the consciousness of belonging to an economy of the gift that precedes every human movement of charity.
>
> (212)

See also Paul Ricoeur, *Lectures on Ideology and Utopia*, ed. George H. Taylor (New York: Columbia University Press, 1986), 321.

11 Paul Tillich, *Theology of Culture*, ed. Robert C. Kimball (Oxford: Oxford University Press, 1959), 5.

12 See Ricoeur, *Lectures on Ideology and Utopia*, 261.

13 See Paul Ricoeur, "Religion and Symbolic Violence," trans. James Williams, *Contagion: Journal of Violence, Mimesis, and Culture* 6, (1999):

> The fundamental feelings associated with religious experience are nowhere visible in their naked immediacy for they are already experienced within the parameters of a given language, which is itself the product of a long cultural tradition. To complicate matters further, each religion is itself like a language that possesses its own customs and has its own history. Religion uses its own special vernacular that preconfigures the way in which members of that religion experience the world.
>
> (4)

14 Marianne Moyaert, "From Religious Violence to Interreligious Hospitality," in *Paul Ricoeur in the Age of Hermeneutical Reason: Poetics, Praxis, and Critique*, ed. Roger W. H. Savage (Lanham, MD: Lexington Books, 2015), 165.
15 Ricoeur, *Figuring the Sacred*, 47. This concern "speaks of the laying hold of the necessary and unique thing from whose basis I orient myself in all my choices" (47; see 65).
16 Ricoeur, *Figuring the Sacred*, 46. Faith is thus the "limit of all hermeneutics and the nonhermeneutical origin of interpretation." See 49 ff.
17 Jaroslav Pelikan, *Bach among the Theologians* (Eugene, OR: Wipf and Stock, 1986), 136; see John Butt, "Emotion in the German Lutheran Baroque and the Development of Subjective Time Consciousness," *Music Analysis* 29, no. 1–3 (2010). In his remarks on Heinrich Schütz's compositions, Butt maintains that Schutz's "music achieves at least some of its effect through its own self-sufficiency; if it were entirely subservient to the text, it would surely become indispensable to some extent" (24).
18 Pelikan, *Bach*, 135. Pelikan rightly points out that such an interpretation falls prey to the fallacy of attributing a work's meaning to one that the composer presumably intended. Nevertheless, "Christian artists have felt able to claim that they were not dealing only with the finite and the temporal" (122). For the conviction that "all beauty, including 'secular' beauty, was sacred because God was one, both Creator and Redeemer" (139) justifies the creation of decidedly secular works.
19 Immanuel Kant, *Critique of Judgment*, trans. Werner S. Pluhar (Indianapolis, IN: Hackett Publishing Co., 1987), 98; original emphasis.
20 Kant, *Critique of Judgment*, 106.
21 Ibid., 114.
22 Pelikan, *Bach*, 61; see Paul Ricoeur, *Freedom and Nature: The Voluntary and the Involuntary,* trans. Erazim V. Kohák (Evanston, IL: Northwestern University Press, 1966). The anxiety of contingency that, "illuminated by the knowledge of death, lends its sting and grief to the thought to death … [so that] death is my death and that my death is anxiety" (Ricoeur, *Freedom and Nature*, 462) evinces the existential ground of this Pietistic sentiment.
23 Ricoeur, *Figuring the Sacred*, 124.
24 Ibid., 123.
25 Ibid.
26 Emmanuel Levinas, *Time and the Other*, trans. Richard A. Cohen (Pittsburgh, PA: Duquesne University Press, 1987), 52. The

> "I" is not initially an existent but a mode of existing itself, that properly speaking does not exist. To be sure, the present and the "I" turn into existents, and one can form them into a time, so that they have time like an existent.… [But then] [i]t is no longer time in its schematic function between existing and existent, time as the pure event of hypostasis.
>
> (53–54)

27 Ricoeur, *Figuring the Sacred*, 123.
28 Levinas, *Time and the Other*, 54.
29 Paul Ricoeur, *Time and Narrative*, vol. 1, trans. Kathleen McLaughlin and David Pellauer (Chicago, IL: University of Chicago Press, 1984), 26.
30 Karol Berger, *Bach's Cycle, Mozart's Arrow: An Essay on the Origins of Musical Modernity* (Berkeley, CA: University of California Press, 2007), 98.
31 Berger, *Bach's Cycle*, 59.
32 Ibid., 98.
33 Pelikan, *Bach among the Theologians*, 62.
34 Ricoeur, *Figuring the Sacred*, 43.

35 See Ricoeur, *Figuring the Sacred*. The

> kind of height that reflexive philosophy confesses indicates the superiority of an originary affirmation that is not numerically distinct from empirical consciousness. The adequation of this empirical consciousness with the spiritual act that constitutes it even defines the itinerary for the liberation of freedom.
> (113)

See also Ricoeur, *Essays on Biblical Interpretation*, ed. Lewis Seymour Mudge (Philadelphia, PA: Fortress Press, 1980). The originary affirmation that Ricoeur opposes to the claim to absolute knowledge opens the way to the reciprocal relation between the "promotion of consciousness and the recognition of the absolute in its signs" (151). Ricoeur points to three indices that mark out the impossibility of any absolute knowledge. The first index of this impossibility "expresses the impotence of fixing the criteriology of the divine in a closed system" (152). The second "expresses the impotence of consciousness to bring all the signs [of the absolute] together in a coherent whole" (152). The third "expresses the impotence of identifying absolute reflection and absolute testimony itself raised to the rank of proof in the grand trial of meaning" (152–153).

36 St. Bishop of Hippo Augustine, *Confessions*, trans. Richard Sydney Pine-Coffin (Middlesex: Penguin Books, 1961), bk. 10 (§33), 238.

37 Augustine, *Confessions*, 238.

38 Ibid., 239.

39 Ibid.

40 Ibid.

41 Ibid., 238.

42 Ibid.

43 Ibid.

44 Ibid., 239.

45 See Charles T. Mathewes, "Augustinian Anthropology: Interior intimo meo," *Journal of Religious Ethics* 27, no. 2 (1999). Mathewes argues that for Augustine, we are free only "when we love (and thus will) a certain end" (205). According to this Augustinian view, the "self is most free when it is determined by God; thus, true freedom is found not only *through* but even *in* the divine imposition of prevenient grace" (195–196; original emphasis).

46 Augustine, *Confessions*, 238.

47 Paul Ricoeur, *Fallible Man*, trans. Charles A. Kelbley (New York: Fordham University Press, 1986), 86.

48 Ricoeur, *Freedom and Nature*, 258. Emotion "consists not only in affective, but also in motive anticipation of goods and evils" (256) that valorize the ends to be pursued. Emotion is accordingly rarely cerebral. Rather, its corporeal disturbances overcome the will's inertia through inciting desire. Desire is also an emotion. Its realm "is as broad as the field of human values which are not only vital, but also social, intellectual, moral, and spiritual" (264). Desire thus "belongs to the body in terms of visceral intensity and muscular alerting which orchestrate, often most covertly, the most subtle movement of the soul" (264). Desire is both motive and motor aimed at fulfilling the need for an intended good. Moved by desire, the body "is the fullness and flesh of anticipation itself" (257). Here, Ricoeur says,

> we have reached the highest point of the corporeal involuntary: desire is a type of spirit of adventure which rises from the body to willing, and which is the reason why willing would have little efficacy if it were not sharpened first of all by the prodding of desire.
> (265)

49 Ricoeur, *Fallible Man*, 84; original emphases. Cf. Will Schrimshaw, "Non-Cochlear Sound: On Affect and Exteriority," in *Sound, Music, Affect: Theorizing Sonic Experience*, ed. Marie Thompson and Ian Biddle (New York: Bloomsbury, 2013). Schrimshaw identifies the affective excesses that he argues are excluded by semiotic and semantic interpretations of music's meaning with the affirmation of the metaphysics of the subject's interiority and the "ideology of individualism" (28). Hence, for him, these affective excesses delimit the boundaries of discourses that claim to decipher or decode music's meaning.

50 Ricoeur, *Fallible Man*, 89; see Paul Ricoeur, *The Rule of Metaphor: Multi-Disciplinary Studies of the Creation of Meaning in Language*, trans. Robert Czerny et al. (Toronto: University of Toronto Press, 1977), 245.

51 Hans-Georg Gadamer, *Dialogue and Dialectic: Eight Hermeneutical Studies on Plato*, trans. P. Christopher Smith (New Haven, CT: Yale University Press, 1980), 66. Praise not only gives those it binds together a common obligation; in praising someone or something, we also avow our commitment to that which we praise. For "[h]e who praises addresses both himself, … the one before whom he praises … and in a certain sense even the one who is being praised."

52 Paul Ricoeur, *The Course of Recognition*, trans. David Pellauer (Cambridge, MA: Harvard University Press, 2005), 222. Paul's hymn to love in Corinthians 1:13 is for Ricoeur the paradigm of the discourse of *agape*. What is more, "[a]gape takes another step in the direction of justice by assuming the form of the imperative.… We could even dare to speak here of a poetic usage of the imperative, close to that of the hymn and the benediction."

53 Ricoeur, *The Course of Recognition*, 222.

54 Ricoeur, *Fallible Man*, 105–106; see Ricoeur, *Freedom and Nature*, 260 ff.

55 Ricoeur, *Fallible Man*, 99.

56 Ibid., 81–82.

57 Ricoeur, *Fallible Man*, 82. According to Ricoeur, Plato attempts to situate *θυμός* "between ἐπιθυμία, sensuous desire, and reason, whose specific desire is *ἔρως*. Thus the inner conflict of human desire reaches its climax in *θυμός*" (92).

58 Ricoeur, *Fallible Man*, 102.

59 Ibid.

60 Ricoeur, *Fallible Man*, 102. Reason demands this totality as the thought of the unconditioned. In reason, Ricoeur explains, "I 'demand' what I 'pursue' in action and that to which I 'aspire' in feeling." Happiness therefore "has the same breadth as reason; we are capable of happiness because reason 'demands the absolute totality of conditions for a given conditional thing'" (Kant; quoted in Ricoeur, *Fallible Man*, 102). Feeling is accordingly "more than the identity of existence and reason in the person; it is the very belonging of existence to the being whose thinking is reason" (103).

61 See my "Exemplarity and the 'Law of Superabundance,'" *Budhi: A Journal of Ideas and Culture* 20, no. 2 (2016).

62 Pelikan, *Bach*, 90. Reimarus (1694–1768) was one of Bach's contemporaries whose book *Apologia for the Rational Worshipers of God* was published posthumously.

63 Pelikan, *Bach*, 96.

64 Ibid., 78.

65 Pelikan, *Bach*, 78. Luther's Christocentric existentialism, the Reformation heritage, and German Pietism all figure in this devoutly sacred work.

66 Pelikan, *Bach*, 95; see 114–115. See also Berger, *Bach's Cycle*, 45 ff. Berger provides a detailed account of the genesis of the libretto based on the text by Christian Friedrich Henrici, who was the librettist for the *St. Matthew Passion* and who published under the pen name Picander.

67 Pelikan, *Bach*, 95.
68 Pelikan, *Bach*, 98. Pelikan points out that it

> has become customary for theologians to contrast the idea of atonement as satisfaction à la Anselm and the idea of atonement as victory à la Luther and the Greek fathers, usually at the expense of the former. By solemnizing the former in his *Saint Matthew Passion* and celebrating the latter in his *Saint John Passion*, Bach demonstrated once again his refusal to choose from among the alternatives that had equally legitimate authority in his tradition. (115)

The *St. Matthew Passion* was deeply influenced by Pietistic theology. Pelikan accordingly notes that "attempts by Orthodox Lutheran confessionalists, in his time or in ours, to lay claim to Bach as a member of their theological party will shatter on the texts of the cantatas and the *Passions*" (57). See Berger, *Bach's Cycle*. Berger argues that the da capo form that Bach employs in his *St. Matthew Passion* acquires its theological significance as the emblem of eternity. He accordingly identifies the "cyclical time" of the da capo form with time's annihilation, which he contrasts with the "arrow of time" (linear time). See also Victor Lederer, *Bach's St. Matthew Passion: A Closer Look* (New York: Continuum, 2008). According to Lederer, Bach employed the compositional aesthetics of affects to "express an intense, nameable emotion" (3). From this vantage point, the opening lament's "mood of somber exaltation" is attributed to Bach's reliance upon musical-rhetorical figures such as *pathopoeia* and *hypostasis*. Lederer consequently maintains that Bach and his librettist sought to portray the "soul reflecting with feeling" (13) on Christ's suffering, God's mercy, and on divine glory. See David and Mendel, *The Bach Reader*, 34; Dietrich Bartel, *Musical Poetica: Musical-Rhetorical Figures in German Baroque Music* (Lincoln, NE: University of Nebraska Press, 1997). Bartel defines *pathopoeia* as a "musical passage which seeks to arouse a passionate affection through chromaticism or some other means (359)," and *hypostasis* as one that "vividly and realistically illustrate[s] a thought or image found in the text" (307). See also Walter Emery, "Bach's Symbolic Language," *Music and Letters* 30, no. 4 (1949). The symbolic language that Emery claims Bach employed draws on the conventions set out by the doctrine of affects. Hence for Emery, Bach "was not content to write joyful music for Easter; he also symbolized the Resurrection by making his music rise. He was not content to write sorrowful music for Good Friday; he also symbolized sorrow by slow chromatic scales" (346). Emery also points out that the three organ chorale preludes on "Dies sind [die heil'gen zehn Gebot]," for example, "show that Bach did not take symbolism as seriously as his commentators do" (347). See Chapter 1 for a further discussion of the doctrine of affects.
69 Hans-George Gadamer, *The Relevance of the Beautiful and Other Essays*, ed. Robert Bernasconi, trans. Nicholas Walker (Cambridge: Cambridge University Press, 1986), 162; see Paul Ricoeur, *Oneself as Another*, trans. Kathleen Blamey (Chicago, IL: University of Chicago Press, 1992). Ricoeur points out that the "aptitude for being-affected through the mode of injunction … [seems] to constitute the condition of the possibility of the empirical phenomenon of identification" (354).
70 Eric Chafe, *Tonal Allegory in the Vocal Music of J. S. Bach* (Berkeley, CA: University of California Press, 1991), 391.
71 Chafe, *Tonal Allegory*, 15.
72 Ibid..
73 Begbie, *Music, Modernity, and God*, 216.
74 Ibid., 69.

75 Begbie, *Music, Modernity and God.* Begbie comments that Butt regards "Bach as 'standing on the cusp of musical modernity,' *combing* repetitive symmetrical components and directionality …, thus mingling pre-modern and modern sensibilities" (68; original emphasis).

76 Jeremy S. Begbie, *Theology, Music and Time* (Cambridge: Cambridge University Press, 2000), 63. According to the myth of passage, which rests on the cosmological conception of time, the present lacks any temporal extension. In his discussion of Augustine, Begbie points out how the ontological deficiency brought to the fore by Augustine's meditations on time and eternity is indicative of our experience of time's dispersion across the present, past, and future. Yet by privileging the question of time's measurement over that of the being and the non-being of time, he undercuts the force of the enigma that springs from tying the soul's distention with respect to time's measurement to the slippage that finds its way into the threefold present (75 ff.; see Chapter 2).

77 See Begbie, *Theology, Music and Time*, 53 ff. It is difficult to understand how a sense of eternity could be evoked by arresting the directed movement of tension and release apart from the operation through which the listener draws a figure from a succession or events. See Chapters 1 and 2.

78 Pelikan, *Bach*, 91. The text of the stanza is

Aus Liebe will mein Heiland sterben,
Von einer Sünde weiß er nichts,
Daß das ewige Verderben
Und die Strafe des Gerichts
Nicht auf meiner Seele bliebe.

For love my Savior now is dying,
Of sin and guilt He knoweth nought,
So eternal desolation
And the sinner's righteous doom
Shall not rest upon my spirit.

(J. S. Bach, *Matthäus-Passion* [*The Passion according to St. Matthew*], ed. Max Schneider (London: Ernst Eulenburg Limited, 1935). Reprint of the 1929 edition by George Schumann, with additional edits by Max Schneider in 1935.) All texts and translations quoted from the *St. Matthew Passion* are from this edition.

79 Pelikan, *Bach*, 100; see 91.

80 Pelikan, *Bach*, 100.

81 Ibid.

82 Pelikan, *Bach*, 12.

83 Karl Barth, *Church Dogmatics*, vol. IV of *The Doctrine of Reconciliation*, trans. Geoffrey W. Bromiley, ed. Geoffrey W. Bromiley and Thomas F. Torrance (London: T. & T. Clarke International, [1956] 2004), 252; quoted in David Hill, "The Time of the Sign: 'O Haupt voll Blut und Wunden' in Bach's St. Matthew Passion," *The Journal of Musicology* 14, no. 4 (1996): 516:

> We are not disputing the purely musical greatness of this work. But it also purports to be an exposition of Chapters 26–27 of Matthew's Gospel. And as such it can only confuse those who hear it…. It is neither determined nor delimited by the Easter message.
>
> (Barth, quoted in Hill, 515–516).

84 Pelikan, *Bach*, 14.

85 Ibid., 70.

86 Pelikan, *Bach*, 83; see 82; 84. See also Hill, "The Time of the Sign." Hill points out that the tune was composed by Hans Leo Hassle (515).

87 The text of the fifth stanza is

Erkenne mich, mein Hüter,
Mein Hirte, nimm mich an,
Von dir, Quell aller Güter
Ist mir viel Gut's getan.
Dein Mund hat mich gelabet
Mit Milch und süßer Kost,
Dein Geist hat mich begabet
Mit manchee Himmelslust.

Receive me, my Redeemer,
My Shepherd, make me Thine.
Of every good the fountain,
Thou art the spring of mine.
How oft Thy words have fed me

88 The text of the sixth stanza is

Ich will hier bei dir stehen;
Verachte mich doch nicht!
Von dir will ich nicht gehen,
Wenn dir dein Herze bricht.
Wann dein Herz wird erblassen
Im letzten Todesstoß,
Alsdenn will ich die fassen
In meinen Arm und Schloß.

Here would I stand beside Thee;
Lord, bid me not depart!
From Thee I will not sever,
Though breaks Thy loving heart,
When bitter pain shall hold Thee
In agony opprest,
Then, will I enfold Thee
Within my loving breast.

89 The text of the first stanza is

Befiehl du deine Wege
Und was dein Herze kränkt
Der allertreusten Pflege deß,
Der den Himmel lenkt;
Der Wolken, Luft und Winden
Gibt Wege, Lauf und Bahn,
Der wird auch Webe finden,
Da dein Fuß gehen kann.

Commit thy way to Jesus,
Thy burdens and thy cares;
He from them all releases,
He all thy sorrow shares.
He gives the winds their courses,
And bounds the ocean's shore,
He suffers not temptation
To rise beyond thy power.

90 The texts of the two stanzas are

O Haupt voll Blut und Wunden
Voll Schmerz und voller Hohn!
O Haupt, zu Spott gebunden
Mit einer Dornenkron'!
O Haupt, sonst schön gezieret
Mit höchster Ehr' und Zier,
Jetzt aber hoch schimpfieret:
Gegrüßet seist dur mir!

Du edles Angesichte,
Vor dem sonst schrickt und scheut
Das große Weltgerichte,
Wie bist du so bespeit!
Wie bist do so erbleichet,
Wer hat dein Augenlicht,
Dem sonsts kein Licht nicht gliechet,
So schändlich zugericht't?

O bleeding head sore wounded,
Defiled and put to scorn!
O Kingly Head surrounded
With mocking crown of thorns!
What sorrow mars Thy grandeur?
Can death Thy bloom deflower?
O countenance whose splendor
The hosts of heaven adore.

Thy beauty long desired,
Has vanished from our sight.
Thy power is all expired,
And quenched the Light of Light.
Ah me! For whom Thou diest,
Hide not so far Thy grace;
Show me, O Love most highest,
The brightness of Thy face.

91 Pelikan, *Bach*, 86. Pelikan's interpretation of the different keys in which the chorale is set emphasizes their programmatically affective significance.
92 Hill, "Time of the Sign," 532.
93 Hill, "Time of the Sign," 532. From a Schenkerian standpoint, the "fundamental composition is askew because the head of the 3-line [third scale degree in F] (a', upbeat to m. 1) does not appear simultaneously with the tonic note in the bass (m. 1: 4)."
94 The prescient G dominant seventh chord (m 8) prepares the way for the phrase's cadence on G minor (submediant of F), which in turn is the dominant preparation for the final phrase, which cadences on F. The opening phrase in D minor, which functions as a tonic substitute, thus sets the tone of the chorale by folding this seemingly deceptive beginning into its configuration as a whole. The prolongation of D minor in the second phrase, and the repetition of the two opening phrases, confirms this opening phrase's submediant's function as a tonic substitute. Following the repeat of the opening two phrases, the fifth phrase (like the first) begins on the submediant and cadences on the tonic (F major).

At the beginning of the sixth phrase the unaccented passing e♭ transforms the tonic (F major) into a secondary dominant that subsequently resolves deceptively, thereby precipitating the movement to the supertonic's (G minor) own dominant (D major). This supertonic phrase, which also commences with a secondary dominant (V7/V), serves as a preparation for the seventh phrase, which cadences on the dominant (C major). The final phrase confirms the nested modulatory progression of the phrases (||:I—vi—I :||: I—ii—V—I:||) that effectively bring the tonal and affective play set in motion by the chorale's initial D minor harmonization to a close.

95 The text of the final setting is

Wenn ich einmal soll scheiden,
So scheide nicht von mir!
Wenn ich den Tod soll leiden,
So tritt du dann herfür!
Wen mir am allerbängsten
Wird um das Herze sein,
So reiß mich aus den Ängsten
Kraft deiner Angst und Pein!

Be near me, Lord, when dying,
O part not Thou from me!
And to my succour flying,
Come, Lord, and set me free!
And when my heart must languish
In death's last awful throe,
Release me from mine anguish,
By Thine own pain and woe.

96 Pelikan, *Bach*, 87.

97 See Ibid., 138.

98 Wilfrid Mellers, *Bach and the Dance of God* (New York: Oxford University Press, 1981), 70.

99 Ricoeur, *Freedom and Nature*, 260. The sense of awe to which we yield in the face of something wondrous not only colors our temporal experiences, but through this sense of wonder objects touch us as though we encounter them for the first time. Wonder is therefore far from a simple reflex; rather, the impact it has is a "shock of knowledge in a disturbance of the body" (254) awakening our thirst and desire. Thought becomes incarnate in the corporeal disturbances that register this shock of wonder. In the face of what is wondrous, fearful, or awesome, emotions resounding in the body reverberate with values and thus incline the will to act.

4 Hope's despair

In the first decade of the twentieth century, Arnold Schoenberg broke with three hundred years of tradition by renouncing the practice of bringing a work to a close on a tonal center. Since the end of the Renaissance, European art music adhered to the principle that dissonances resolve in order to achieve tonal closure. Schoenberg's renunciation of this long-standing principle not only shattered established conventions, but by emancipating dissonance from the normative requirement that governed music's tonal progress, Schoenberg also abruptly upended the ideal that in the common practice era undergirded tonal music's temporal coherence.

By relating Schoenberg's renunciation of tonal strictures to the crisis brought about by the loss of confidence in the Hegelian philosophy of history, I want from the outset to highlight how the negation of tonal closure in Schoenberg's freely atonal works calls into question the ontotheological presumption of a way of thinking that dares to totalize time. The philosophical temptation to grasp history in its entirety is tantamount to raising thought about time to the level of the absolute. For such a style of thought, time's totalization in the eternal present is the mark of this style of thought's philosophical mastery. Schoenberg's negation of tonality contests this philosophical style of thought's ideal of the oneness of time. As I will explain in due course, the emancipation of dissonance by analogy overturns the hubris of a mode of thinking that claims to comprehend the world-historical spirit's rationally necessary plan through subverting the dialectical identity exemplified by the traditional sonata form.

My investigations in this chapter are governed in part by my critique of the apparent collusion between this philosophical way of thinking and tonal music's teleological cast. In the course of the reflections that follow, I will reiterate how in tonal music the sense of temporally directed progress draws its force from the synthetic operation that schematizes this sense of forward-directed motion by drawing together successive thematic and harmonic components. As I explained in Chapter 1, the sensation of tonal movement is a function of this synthetic operation, to which I also attributed music's affective force. The resolution of dissonance propels this movement forward only to the degree that the succession of individual sonorities contributes to structuring the harmonic progressions of which each is a part. The drive toward resolution, and the play of tension and release, are effects of the manner in which heterogonous elements cohere. Tonality's teleological cast is accordingly a nomic generality abstracted from a temporally dynamic configuration in which the expectation of the dissonance's resolution figures.

The guiding question of this chapter, then, is "How do works in which the pull of tonal resolution no longer holds sway bear on the way that we understand the force of the present?" By setting the force of the present against the Hegelian temptation to totalize time, I propose to relate as yet unfulfilled claims to the structure of action. The initiative pursued by Schoenberg in emancipating dissonance from the requirement that it resolve is exemplary in this regard. The step taken by Schoenberg is comparable to overturning a system of philosophical thought for which the whole is the mark of its truth. The break with the conceit that this system is identical with reason as it manifests itself in history reserves a place for hope. It remains to be seen whether Schoenberg's negation of a nomic principle that effectively guaranteed a sense of temporal continuity necessarily replaces the force of reason in the Hegelian system with a negative ontotheology. At the very least, the irruption of unresolved dissonances in Schoenberg's freely atonal works calls into question the possibility of continuing to make history at all.

Negations

The notion that the negation of tonality's principal mainstay jeopardizes the continuity of a series of developments that mark out the course of Western art music's inexorable advance admittedly flies in the face of Schoenberg's view of his own historical importance. By refusing to force his music into the Procrustean bed of traditional harmonic relations, he and many musicologists who have commented on his work regard his stylistic, innovative break with tonal precepts as a logically necessary step. Deprived of the support of harmonic progressions that would justify concluding a passage or a work with a traditional cadential ending, tonal closure was stripped of its legitimacy. In his *Second String Quartet*, Op. 10, Schoenberg consequently explains that he found that "occasional returns to such tonal triads as represent a key"[1] no longer served as a viable counterforce to the overwhelming number of dissonances spawned by individual parts of the first and second movements. Accordingly, his *Two Ballads*, Op. 12 marks the transition to a second style period, in which he rejected the idea that music's expressions should be governed by a tonal center. Schoenberg saw himself as the first among his contemporaries to take this historically decisive step. In response to these new compositional demands, he placed himself at the forefront of Western art music's inexorable advance in the face of the faltering credibility of traditional tonal procedures.[2]

My claim that Schoenberg's response to the dilemma he faced bears a resemblance to the crisis brought about a failing confidence in the ideal of historical progress is decidedly at odds with Schoenberg's view of his own significance. Schoenberg was convinced that since the rhythmic and motivic materials with which he was working did not support their tonal resolution, the traditional way of concluding a work could no longer be justified compositionally.[3] For him, the contradiction between the way that these materials work together to express the ideas to which they give a figure and a body and the nomic principle of tonal closure was inescapable. The decisive step that he took is a testament to the rigor and integrity with which he pursed his compositional métier. Accordingly, he renounced tonality's governing role in order to ensure that music's expression was not fettered by extraneously imposed compositional rules or laws.

Schoenberg's response to the dilemma engendered by the proliferation of chromatic dissonances in late Romantic music, however, can only be reduced to a causal connection between them at the cost of eclipsing the critical historical import of his aesthetic innovation. Subordinating the initiative that he took in renouncing functional tonal relations as the basis of a work's organization to the rule of a covering law model such as Karl Hempel advanced falls prey to the historicist thesis that the object of history is to discover causal relations among logically ordered sequences of events. Leo Treitler reminds us that situating events in a historical chain accedes to the standard of truth that is derived from this thesis.[4] Schoenberg's venture in embarking on this new path could have been regarded as necessary—even if it was not regarded universally in a positive light—only if, given certain conditions, a consequent event inevitably follows. Conversely, renouncing the historicist thesis opens the way to understanding the historical significance of individual works in the context of an interpretation of the response that each work provides to a question or problem for which it provides an answer.

The aporia that Dahlhaus identifies when he sets the requirements of music historiography against the strong claim to music's aesthetic autonomy places the hermeneutical value of this tack in relief. This aporia highlights the impasse that arises when music history is set against the analysis of individual works. On the one hand, sacrificing a sense of historical continuity to individual analyses dooms music history by isolating singular works from the history of which they are a part. On the other hand, reverting to a history of social or intellectual developments in order to cobble together a cohesive narrative fails as a history of music.[5] Dahlhaus is right to insist that art history's vindication rests on an interpretation in which the historian grasps the work's historical character in terms of its aesthetic constitution. In the final chapter, I will show how the "rule" summoned by a work relates to the work's singular response to a question or crisis as apprehended by the composer or artist. The fit of the work, I will say then, exemplifies the "rule" that is the spring of the work's claim to truth. As such, creative innovations stand as responses to problems or crises arising from a history of prior discoveries and inventions to which these innovations contribute and of which they are a part.

By situating Schoenberg's negation of tonality within this dialectical interplay of novel inventions and a history of prior developments (an interplay that constitutes the dialectic of tradition), I mean to recover the force of his initiative. The reasons that he broke with tonal strictures are patently obvious. Treating this initiative as though it were logically and historically inevitable, however, occludes both its creative and critical significance, not to mention the affective and ontological vehemence of the works in question. By insisting that incremental transformations in Schoenberg's musical language in the *Four Songs*, Op. 2 (1989) and *Verklärte Nacht* (1899) led to Schoenberg's break with the tonal tradition in August 1909, Ethan Haimo ratifies the convention of identifying Schoenberg's break with tonality as a natural progression. According to Haimo, Schoenberg consolidated this break with *Gurre-Lieder*, *Erwartung*, and *Pierrot Lunaire* during the period 1909 to 1911; the stylistic advances that led to the dissonance's emancipation therefore could be regarded as the "product of nearly a millennium in the evolution of Western music."[6] Following this line of reasoning, Pierre Boulez's radicalization of Schoenberg's twelve-tone method could then be said to represent a further stage of the logic unleashed by efforts to domesticate the dissonance's disruptive force.[7]

This evolutionist perspective clearly trades on the tonal tradition's ontotheological underpinnings. Kerman rightly emphasizes that the principle of organicism constituted the Western art music's tradition's ideological core. This tradition's foundation did not depend upon the "triad or tonality, as [Heinrich] Schenker and [Donald Francis] Tovey believed,"[8] but instead rested on its affinity with an ontotheological view of history that legitimated the value placed on a work's internal system of organization. The idea of progress that was rooted in a "secular formation of a concept which originally was a theological one"[9] redoubles the resemblance between tonality's teleological characterization and a theodicy that attempts to justify the inscrutability of the evil that threatens to destroy history's rational course. If tonality's unifying power could be seen as comparable to that of reason, it was only to the extent that this unifying power drew its force from Romantic philosophy's understanding of itself "as the place where reason, as the absolute power of unification, entered upon the scene."[10] However unintended this consequence of Romantic philosophy might be, music's metaphysical elevation ratified the comparison of tonality's unifying power with reason's power of comprehension.

The ideology that Kerman identifies with the ideal of a work's organic development accordingly took root in the soil of Romantic philosophy. Just as "Geist's [Spirit's] embodiment in a rational universe … [could be regarded as] not only the fulfillment of his rationally necessary plan, the Idea, but also the paradigm expression of what that purpose is,"[11] a tonal work's temporally unfolding totality could be interpreted as the expression of an idea the development of which is guided by preternatural forces. Schenker's conviction that all great music blossomed organically paved the way for a method of analysis that justified this ideal, the origins of which Kerman attributes to Johann Nicolaus Forkel's 1882 biography of Johann Sebastian Bach.[12] If for Hegel the task of philosophy was to comprehend "the spirit of the world as it reveals itself through the human consciousness,"[13] for Schenker the goal of analysis was to make explicit the "secret perception of the interaction of origin, development, and present"[14] that constituted the soul of a work's inner logic. Schenker's stated objective—to cultivate the capacity for hearing music "in the spirit of its coherence"[15]—drew its power from an analytic procedure that abstracted a fundamental linear structure from a work's surface features. Consequently, his theory of structural prolongation sought to justify methodologically tonal music's superior aesthetic value.

Schenkerian reductions bring to the fore the contamination of the tonal tradition's ontotheological justification with an evolutionary perspective. Although both the tonal tradition's ontotheological characterization and the notion of its evolving course can be construed teleologically by identifying each with some standard of progress, they differ in one important respect. The model of Western art music's historical evolution depends upon one transferred from biology and zoology to the field of human praxis. In contrast, tonal music's ontotheological value draws support from the presumed demonstration of reason's operative role vis-à-vis the totality of a work. Tonal closure provides a kind of Hegelian guarantee that the work in its totality is the emblem of reason's supreme power. Only a philosopher who has "thought through the conditions that make freedom both rational and real"[16] in Spirit's process of self-realization could equate the idea of freedom with history's unity as a whole. By analogy, a tonal work's integral unity is proof of the philosopher's conviction that reason creates the conditions for its own realization, which finds its justification in the system that confirms this conviction. The Goethean *Urpflanze* from which Beethoven's *Fifth Symphony* blossoms, for Hoffman, folds together the metaphor of organic growth and development with this idea of freedom. The aesthetic freedom won through music's social emancipation could therefore offer safe haven to the human spirit only by reason of tonal music's supposedly inherent value.[17]

It is hard to imagine that Schoenberg could safeguard tonality's ontotheological value in the face of the crisis precipitated by *Verklärte Nacht*'s chromatic saturation. If, as Kerman claims, Schoenberg's decisive insight was to preserve the ideal of a work's organic unity while negating tonality's unifying force, works such as *Erwartung* or *Pierrot Lunaire* dispense with any nomic principle that would guarantee their temporal coherence and continuity in the absence of tonal closure. Susan McClary is partially right when she identifies Schoenberg's transgression of tonal conventions in *Erwartung* with the liberation of dissonance's disruptive thrust. For Schoenberg subverts tonal music's ostensibly rational character by refusing to resolve dissonance in accordance with established procedures. This rational character owes its credibility to tonality's ontotheological characterization.

By insisting that Schoenberg escaped tonality's narrative grip in *Erwartung* by means of the metaphorical surrogate of the female lunatic, McClary consigns dissonance's temporally dynamic quality to a semiotics of gender. Hence, for her, Schoenberg's transgression of traditional tonal boundaries at the same time "betrays his inability to dismiss or transcend traditional binarisms and their gendered associations."[18]

By setting the excesses of dissonances in *Erwartung* against a tonal frame that presumably can no longer contain these excesses, McClary effectively strips dissonance of its phenomenological meaning. Hence, for her, these excesses are marked structurally or semiotically as "feminine" by this tonal frame, which also places this "feminine Other" on display.[19] These "dissonant" excesses are accordingly signs of difference, which is subjugated by, or subordinate to, the hegemonic forces of a patriarchal political agenda. Reason and "feminine" hysteria are semiotically marked by their respective positions in the tonal narrative that McClary places at the center of her critique of absolute music's enactment of this agenda. Mapping these differences of position on to socially codified differences between male and female and between reason and unreason thus constitutes the formal *modus operandi* for deconstructing absolute music's presumed aesthetic autonomy.

By identifying reason with a tonal narrative that frames the "feminine Other" in order to contain it, McClary's deconstructions clearly trade on dissonance's phenomenological significance. Although music differs from narrative in important respects (as I explained in the Introduction), the parallel between tonality's ontotheological justification and "grand historical narratives" in the style of a Hegelian philosophy of history supports a critique of the identification of reason with the cultural ideal of a work's organically unfolding totality. Conversely, if we continue to assert that a tonal work's logic inheres in the way that the play of parts and whole constitute an integral totality, then we must also admit that dissonance acquires its meaning only in terms of its power of negation. Philosophically speaking, this power of negation is the driving force of thought. Hegel could claim that "The True is the whole"[20] only because in his system Spirit comes to know itself as absolute by embodying itself through the negation of the negations through which Spirit equally expresses itself. Adorno's conviction that the "whole is the false"[21] leads to a totalizing critique that loses itself in its relentlessly negative dialectical strategy through reversing Hegel's dictum, since for Hegel the "whole is nothing other than the essence consummating itself through its development."[22] The tonal dynamics of the traditional sonata form thus assume the guise of a cipher for Spirit's fulfillment of its rationally necessary plan by reason of dissonance's sublation. (It is perhaps worth mentioning that Beethoven premiered the *Eroica Symphony* just two years before Hegel published his *Phenomenology of Spirit*.) Within this ontotheological context, dissonance is the negative emblem of a demand for totality that paradoxically links a philosophical style of thought that purports to comprehend the whole to a sociomusical-theoretical discourse rooted in a theodicy of dissonance's disruptive thrust.

Before I go on to discuss how Schoenberg's conception of musical ideas lends itself to a hermeneutical reinterpretation, I want to emphasize once more how Schoenberg's rejection of tonal closure's nomic role calls into question the modernist presumption of the continuity of history and time. By claiming that Schoenberg's negation of tonality in his freely atonal works is comparable to the Hegelian philosophy's loss of credibility, I am by no means suggesting that he renounces the modernist ideal of historical progress. On the contrary, Schoenberg clearly placed himself at the forefront of the Western art music tradition's advance, despite his ostensible protestation that he was no Bach.[23] By the same token, Schoenberg, like Adorno, insisted that whatever materials it might employ, art's desire for dissonance, which society's hegemonic self-affirmation suppresses in monopolizing art's aesthetic semblance, is the spring of its truth. Art's protest against socially frozen habits of thought gains its practical foothold in the ways that individual works shatter reality's ideologically congealed systems by means of their imaginative explorations of new ways of thinking, feeling, and acting, as I will argue more fully in the final chapter. Sociological readings of modern art's and music's disruptive force thus take hold only in the hermeneutical soil of each work's worlding power (see Chapter 7). If by placing the real into suspense a work such as *Erwartung* subverts established social conventions and norms, it is only to the extent that the note of discord sounded by *Erwartung* is one that the work also sets at the heart of the real. Dissonance in modern music and art is accordingly more than just a sign of music's and art's protest against ideologically reified regimes of power and knowledge. As the emblem of the demand for reconciliation, the dissonance unleashed by Schoenberg's refusal to adhere to tonal dictates by analogy consequently sets the Hegelian philosophy of history's loss of credibility, together with its negative inverse, against the logic of hope.

Musical ideas

Schoenberg's negation of tonal music's nomic schemata immediately raises the question as to how renouncing tonal closure intensifies his music's expressive force. On the one hand, Schoenberg rejected tonality's governing role in order to meet the expressive demands of *Gurre-Lieder* and *Erwartung*. On the other hand, the ideas voiced by these works take shape without the support of any formal overarching system. It is as if in the quest to intensify music's expression, Schoenberg threw off the assurance of temporal continuity afforded by tonality's presumed ontotheological character in order to fulfill the aesthetic requirements of his artistic vision.

Casting off this assurance might reasonably seem to have spawned a sense of anxiety that registers the loss of confidence in social and political institutions at the end of the nineteenth century. Carl Schorske emphasizes that the *fin de siècle* aesthetic movement with which Schoenberg was deeply involved espoused the prevailing idea that "traditional coordinates of ordered time and space were losing their reliability, [and] perhaps even their truth."[24] Aesthetic attacks on the disintegrating authority of good society's decaying culture explored the disordered worlds of the psyche, the reaches of erotic decadence, and the era's pessimistic weight. Suspicions nurtured by Nietzsche, by psychoanalysis, and by ideology critique, along with the "era's great romancers,"[25] fueled this descent into deeper doubt. The idealistic concept of Spirit's self-realization, which Gadamer remarks "had not moved beyond variations on the idealist synthesis of Goethe's time,"[26] was incapable of withstanding the pressures of modernity's scientifically shaped cultural consciousness. Europe's optimism regarding modern progress and bourgeois cultural ideals could not survive the catastrophic onslaught of the First World War's technological slaughter.[27] For Gadamer, this technological slaughter's universal reach has strangely unified the age that it ushered in under the aegis of the crisis of consciousness ignited by the twentieth-century's world wars.

This crisis of a historical consciousness schooled in modernist ideals of progress offers a broader context for the initiative taken by Schoenberg when he refused to force his music into rigidified schemata. The rupture in the temporal continuity of past and future fuels this crisis by laying bare the gap that Hannah Arendt locates between the future and the past. This gap in the continuity of historical time sets off the present as the moment of decision with respect to the initiatives that we take. This gap, Arendt remarks, is not a modern phenomenon insofar as the present is the time in which we intervene in the world's course. Rather, for her, the troubling problem is that we seem to be neither prepared nor equipped for the activity of thinking that she believes is required for political judgment in an era in crisis.[28] The crisis of a historical consciousness for which the assurance of the past, present, and future's temporal continuity has been cast into doubt gives rise to the challenge and the task of preserving the tension between the space of our experiences and the horizons of our expectations. The imperfect mediation called for by this challenge and this task stands as a riposte to the Hegelian temptation to totalize time.[29] As I will explain in due course, the logic of hope reserves a place for this imperfect mediation by inscribing it in the structure of action.

Schoenberg's refusal to force expression into a tonal mold brings to the fore the temporal paradox engendered by absolving dissonance of the requirement that it resolve. Schorske rightly points out that emancipating the dissonance is not the same as constructing another system that would replace tonality (as, for example, in the case of the twelve-tone method). Stripped of its protective wrappers (that is, of its preparation and resolution within a tonal schema), dissonance can no longer be "carefully packed so that it neither suffers nor inflicts damage."[30] Schoenberg's own allusion to tonality's protective frame seems to support an Adornoesque reading that identifies dissonance's emancipation with modern music's expression of suffering.[31] This reading draws dissonance's emblematic negation of harmony into the orbit of a "hope beyond hope." As such, depriving dissonance of tonality's protective frame reveals the distance separating modern art's and music's expression of suffering from the suspended arrival of utopian aspirations.

The notion that dissonance's drive toward resolution is the cause of the feeling of tonal motion has its root in the temporal operation that the suspended arrival of utopian aspirations violently arrests. Placing this feeling of motion under the rule of tonality's ontotheological rubric obfuscates the productive role of the synthetic operation in which dissonance's temporal force figures (see Chapter 1). In accordance with this rubric, tonal closure is said to be a function of tonality's teleological cast. Schoenberg, however, clearly also recognized that the credibility of bringing a work to a close in this way rests on the logic of the musical developments leading up to it. This logic, I also explained in Chapter 1, is an effect of the way that a passage's or a work's constitutive elements cohere. As such, this logic is a function of the fit apprehended by means of the operation through which a configuration is drawn from a succession of heterogeneous elements. Schoenberg's renunciation of a tonal center as a nomic organizational principle places this operation at the forefront of his music's expressive vehemence. Renouncing a tonal center paradoxically intensifies the expressive value of emancipated dissonances by arresting the dissonance's drive toward resolution. Schoenberg's radical negation of tonality's fundamental precept thus lays bare the connection between a work's emphatically temporal expression and its affective vehemence by stripping away this expression's nomic supports.

The phenomenological reading on which the preceding comments regarding the dissonance's temporally expressive value are based is already operative in musico-theoretical perspectives that attribute the birth of tonality to Claudio Monteverdi's use of a dominant seventh chord in his madrigal *Stracciami pur il core* (Book 3 SV 64) or *Cruda Amarilli* (Book 5 SV 94).[32] That tonality owes its significance to the role of the dominant seventh chord underscores how dissonance's negation defines the tonal center as such. Supported by the root progression from the fifth scale degree (dominant) to the first scale degree (tonic), the resolution of the dissonance created by the fourth and seventh scale degrees establishes the tonic as the goal (Example 4.1). Phenomenologically speaking, the drive toward resolution is part of a larger dialectical operation set in motion by dissonance's power of negation. This dialectical operation is borne out by the large-scale function played by the second theme's appearance in the dominant key of the traditional sonata form's exposition, so that the first and second themes' recapitulation in the tonic sublates this large-scale dissonance (see above). The play of tension and release that vests this traditional tonal movement with its expressive character is emphatically temporal. We should therefore ask if, isolated from the temporal operation in which the expectation of dissonance's resolution figures, dissonance and discord would be stripped of their expressive value.

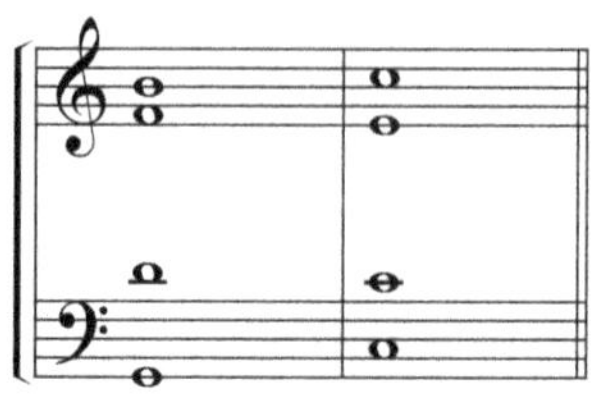

Example 4.1 Dominant 7th Chord.

The anticipation of dissonances' resolution clearly figures prominently in the operation that draws a tonal configuration from a series of heterogeneous elements. Within a tonal context, the apprehension of the fit of tones and harmonies that follow one another in succession is guided by the expectation that dissonant sonorities will resolve. These dissonant sonorities' expressive force depends in large part on this expectation, where the prospect of dissonant tones' resolution can be frustrated or forestalled. Expression, I indicated in Chapter 1 (see also Chapter 7), is a function of the matrix that schematizes the fit of a work's constitutive parts. Expression thus owes its force to the operation that transforms the succession of a passage's or a work's constitutive elements into an affectively meaningful temporal configuration. The play of tension and resolution structured by dissonance's dynamic function accordingly sets the listener's expectations to work. However much nomic principles of tonality guide these expectations, the temporal matrix drawn from successive elements is the emblem of the work's internal logic and the spring of its expressive force.

The virulence of the aporia unleashed by Schoenberg's renunciation of a tonal center's governing role militates against the temptation to confuse the work's expression of its idea with its formal abstraction. By freeing dissonance from the demand that it resolve, Schoenberg not only placed the dissonance's forward-directed thrust radically into suspense, but the negation of tonality's operative system also intensified music's expressive force through casting the figuration of a work's idea back onto the expression of this idea's temporal matrix. Once dissonances were no longer required to resolve, dissonance's expected overcoming (dialectically, the negation of the dissonance's prior negation of a consonant sonority) is effectively stripped of its normative force. Dahlhaus accordingly points out that "[i]nstead of producing musical motion the [now emancipated] dissonance remains, seemingly at least, … without consequences."[33] Hence, the aporia. Since complex sonorities no longer necessarily lead anywhere by reason of some driving force, dissonance's "[i]solation proves to be the reverse of [its] emancipation."[34]

Schoenberg's negation of tonality could be regarded as a step on the way to his invention of the twelve-tone method only because of this aporia. By foregrounding the law that he insisted ensures that dissonances are as comprehensible as consonances, his retrospective justification of the dissonance's emancipation authorized replacing tonality with this method. Schoenberg consequently voided the antithetical relation between consonances and dissonances by maintaining that the difference between them was a function of their relative proximities to a fundamental tone within the overtone series. From this theoretical standpoint, "dissonant tones appear later among the overtones"[35] than do consonant ones. Hence, they can be equally integrated by means of a system that refuses to preserve traditional prejudices regarding dissonance's intelligibility. Schoenberg stresses that a musical "organism lives as vitally in its phrases, rhythms, motifs, and melodies as ever before,"[36] as it does when the drive toward tonal closure animates a work's temporal structuration. To the degree that dissonances are integral to this living organism's expression, their "natural and logical outgrowths"[37] make them as comprehensible as consonances. Thus, by replacing references to a tonal center with a method of composing where every part of a work is derived from "tonal relations in a basic set of twelve tones,"[38] Schoenberg sets out a method of composing that ensured that a work's *Grundgestalt* would be coherent by reason of the fact that every unit would be logically integrated by means of a basic twelve-tone set.

The slippage between the expression of musical ideas and the means of rendering these ideas formally intelligible in terms of a work's internal coherence stems from the temptation to reduce these ideas' temporal figuration to atemporal combinations of elements derived from a basic twelve-tone series. Schoenberg's conviction that there is no form in music apart from the unity of its logic undoubtedly motivated the search for a method that would undergird the comprehensibility of consonances and dissonances independently of a tonal system. Hence, this method's purported necessity is more than just a historical one, since a method in which dissonances are treated as equivalent to consonances must above all else also meet the demand for a seamless unity of music's form and logic. The main advantage of composing using a method in which all twelve tones are related only to each other derives from its singularly comprehensive function. To the extent that this comprehensive function's unifying effect precedes the sense of the fit brought about through the configurating operation that successively draws individual elements together, this comprehensive function supplants the work's temporal matrix as the source of the work's specifically musical logic.

Sidelining the aporia unleashed by renouncing tonal closure in this way in fact supports my thesis that Schoenberg's negation of tonality constitutes an event that is comparable to the event in thinking brought about by attacks on the authority of the Hegelian philosophy of history. By insisting that the "elimination of consonance, of resolution, destroys the basis for expression [and] makes dissonance itself meaningless,"[39] Charles Rosen highlights the degree to which traditional conceptions of musical expression trade on dissonance's relation to a tonal center. At the same time, Rosen attributes the expressive force of Schoenberg's freely atonal works to Schoenberg's "reconstruction of the relation of consonance and dissonance without the use of the perfect triad."[40] Rosen maintains that Schoenberg discovered a way of reformulating this relation within an atonal context. From Rosen's vantage point, it would seem that Schoenberg somehow managed to preserve dissonance's operative temporal function despite having arrested the drive that vested dissonance's tonal resolution with its ontotheological significance in the first place.

The aporia brought into focus by Rosen's insight into the apparent destruction of dissonance's expressive value supersedes the aporia identified by Dahlhaus, where the dissonance's emancipation is the cause of its isolation within the textural fabric of a passage's forward-directed progress. How, we might wonder, do dissonances effectively create a sense of expressive tension once they are no longer required to resolve? Once emancipated, the driving force of dissonance's expected resolution seems permanently to be arrested. What then animates a work's form of expression when there is no apparent distinction between the weight of the tones and combinations of tones that follow one another in succession? Rosen points out that Schoenberg's reformulation of the consonance-dissonance relation is the most impenetrable of all his innovations in this regard. Should we therefore conclude that the expressive vehemence of Schoenberg's freely atonal works is directly proportional to their formal or theoretical opacity?

Schoenberg's own comments on the expression of musical ideas offer a way out of this apparent impasse. He emphasizes how an idea's external appearance—that is, its material expression—can be regarded as adequate only when it meets the demands to which the composer seeks to do justice. The form of expression taken by a work is the measure of the work's adequacy in this regard. As I will explain further in Chapter 7, the "thought" or "idea" to which a work gives a body and a figure in response to the problem or crisis for which the work provides a singular answer is the mark of a work's style. Consequently, the distinction that Schoenberg draws between a style and an idea obscures the relation between them. Schoenberg's claim that "ideas can never perish"[41] is therefore true so long as the history in which scientific and artistic discoveries and innovations figure endures and so long as the experiences occasioned by musical works revivify the ideas that they express.

Schoenberg's conviction that the "totality of a piece"[42] constitutes the idea of a work ultimately underscores the impossibility of reducing this idea's expression to a formal system of musical logic. In a statement that echoes Hanslick's view, Schoenberg tells us that composing "is *thinking in tones and rhythms*."[43] Correlatively, since "[e]very piece of music is the *presentation of a musical idea*,"[44] this idea rises to the level of expression only in accordance with the way that the work itself takes shape as it unfolds. Patricia Carpenter and Severine Neff stress that "Schoenberg's specific theory of music rests on his more general view of art as [the] expression of ideas and the work of art as their realization."[45] For them, Schoenberg's technical concepts—idea

(*Gedanke*), continuing variation, consequence and logic, and the like—are therefore best understood in terms of his guiding notion of the "musical work as [a] concrete presentation of a musical idea."[46] Schoenberg similarly explains that "[e]very tone which is added to a beginning tone makes the meaning of that tone doubtful."[47] For example, the succession of a G major triad by a C major triad could be variously interpreted as a progression in C major, G major, F major, or E minor. This state of unrest, Schoenberg points out, generally increases throughout most of a work. By identifying the work's expression of its idea with the "method by which balance is restored,"[48] he accordingly sets the stage for differentiating between the fit of the work, which I have attributed to the work's temporal configuration, and its formal coherence.

The temporal expression of "thoughts" or "ideas" that I argued earlier are coextensive with the feelings and moods exemplified by a work thus marks a major difference between Schoenberg's negation of tonality's nomic order and the performative contradiction of Adorno's negative dialectical strategy. Adorno, who championed Schoenberg's freely atonal music as a critique of instrumental reason's encroachment on all spheres of life in modern society, regarded Schoenberg's music to be the "music of identity in nonidentity."[49] Conversely, negative dialectics, Adorno tells us, is the "consistent sense of nonidentity"[50] that makes the totality demanded by thought the measure of our finite consciousness's insufficiency. Since for Adorno, "[t]otal contradiction is nothing but the manifest untruth of total identification,"[51] he reverses the Hegelian thesis that gives the identity of reason and the whole that consummates itself pride of place. Hence, Adorno's pessimistic pronouncement: "Dissonance is the truth about harmony."[52] We might legitimately wonder whether works that attain the highest level of form through their internal construction acquire their content by negating meaning. For how, we could ask, does the renunciation of meaning and the refusal to communicate point to a practice from which modern works of music and art abstain, namely "the creation of a just life"[53] that this apparent refusal to communicate places beyond reach?

The notion that music in which emancipated dissonances are given free reign is the negative index of a utopian condition that, so far as modern music and art are concerned, is draped in black is strangely at odds with the demand for expression that motivates the declaration of the so-called state of emergency announced by Schoenberg when he suspended the existing tonal order in the last movement of his *Second String Quartet* Op. 10. Schoenberg's renunciation of tonality no doubt turned out to be decisive in so far as it proved to be historically and stylistically irrevocable.[54] However much Schoenberg's deployment of the technique of developing variation in this new universe seemed to be the compositional analogue of negative dialectics' critical virulence, the expression of ideas turns against the performative contradiction that in the end swallows up this totalizing critique's relentless negativity.[55] The resistance to this performative contradiction that comes from the force of expression in Schoenberg's music leads me to the heart of the question I set out at the beginning of this chapter, namely whether Schoenberg's renunciation of tonality rises to the level of an event comparable to the event in thinking that discrediting the Hegelian philosophy of history brought about. This event in thinking highlights the problematic of history's temporalization. In turn, the question that I posed calls for a further reflection on the emblematic value of the initiative taken by Schoenberg vis-à-vis the logic of hope.

Erwartung

Schoenberg's monodrama *Erwartung* is an explosive testament to the expressive forces unleashed by his renunciation of a tonal center. Set to Marie Pappenheim's text, which Schoenberg adapted for this musical *Angstraum* and which was ostensibly a realist study of feminine hysteria, *Erwartung* explores the innermost recesses of regions of feelings assigned to its solitary figure's tormented soul. The monodrama's solitary figure, who is identified only as Woman, purportedly represents the Viennese upper class of the *fin de siècle*. Pappenheim's *Erwartung* has accordingly been interpreted as a critical exposé of good society's psychic vulnerabilities.[56] On this view, the work's portrayal of a disordered world of feelings and emotions is the index of a radically fragmented reality. From this point of view, the Woman, who is "incapable of ordering her reality, [and hence is] unable to find her way,"[57] personifies the *fin de siècle* crisis of confidence with which Schoenberg in his own way was deeply involved. Schorske points out that "like [Oskar] Kokoschka, Schoenberg took the initial step into a new world of feeling within the framework of lyric poetry, but then quickly broke into the open with a radical turn to theater."[58] Liberated from tonal constraints, this singular exploration of a world of feeling exploits the expressive forces channeled through disparate configurations of sounding events.

While the affinity between *Erwartung* and the crisis of consciousness fueled by the hermeneutics of suspicion is more deeply rooted in the work's radical negation of tonality's ontotheological impulse, the subject matter lends itself to interpreting *Erwartung*'s affective character as the expressive register of its solitary figure's mental states. Joseph Auner remarks that Schoenberg's vision of composition "as the transcription of the constantly changing and irrational unconscious"[59] erases the boundaries between inner psychic impulses and the disordered reality of an alienating world. By abandoning the security of harmonic order, Schoenberg effectively creates a musical portrait of this solitary figure's deranged psychic condition through radically attenuating the work's unifying elements. This affective portrayal of the Woman's psychic distress clearly rests on the work's figuration of changing feelings and fluctuating moods. Through drawing together fragmented remnants of consonances' and dissonances' voided tonal relation, the monodrama evokes the atmospheric world of the Woman's mentally fragile state.

Erwartung is accordingly a musical *tour de force*. If, as Bryan R. Simms suggests, the theory of emotions espoused by Schoenberg is key for understanding *Erwartung*'s experimental language, it is only to the extent that this language is coextensive with the style and manner of the work's expression.[60] Simms emphasizes that Schoenberg's focus on emotions in *Erwartung* is "symptomatic of his broader interest in the human psyche as the source of artist creativity and his own atonal musical language."[61] Hence, in Simms's view, *Erwartung* approaches a peak (on the other side of which lay the beginning of a modern reformulation of a classical conception in *Die glückliche Hand*) through freeing itself from formal principles, archetypes, and conventional motivic procedures. The idea that *Erwartung* registers the incoherence and irrationality of emotions that erupt spontaneously from an unconscious mind seemingly authorizes a reading that identifies feelings and moods expressed by the work with emotions that reflect the Woman's inner physic experiences. Yet as I have pointed out, the world projected by a work is one where the work's mimetic representation of feelings and moods effects the metamorphosis of everyday emotions.

Erwartung's stylistic *tour de force* is thus the spring of its mimetic power. Peter Bürger rightly stresses that the "avant-gardiste work does not negate unity as such"[62] in this regard. Rather, works that in the twentieth century turned against traditional precepts subvert the ideal of the work's organic totality. Fueled by the revolt against the bourgeois cult of *Bildung*, the avant-garde's self-criticism of art's institution as a separate sphere isolated from the material order of life rebels against the principle of a seamless totality of whole and parts. Despite the avant-garde's negation of this emblematic synthesis, the individual elements' emancipation is never totally detached from the work as a whole. On the contrary, these emancipated elements acquire their critical expressive significance only through contributing to the formation of the tensions that inhere in the relation between parts and whole.[63] Hence, the avant-gardiste work can be understood only hermeneutically as a discordant totality. Since the work's unity is no longer predicated on the harmonious integration of individual parts, the work foregrounds fissures and discordances that negate a traditional work's organic integrity. Consequently, the "contradictory relationship of heterogeneous elements"[64] takes shape only by reason of the configurating operation from which these fissures and discordances emerge. Hence, even "where the negation of synthesis becomes a structural principle, it must remain possible to conceive however precious a unity"[65] if the ruptures and interstices articulated by an avant-gardiste are to be seen, heard, or felt at all.

The absence of tonality's nomic generalities is paradoxically the key to *Erwartung*'s success. The work's resistance to its conceptual recuperation either through formal analysis or through psychoanalytic interpretations is indicative of its singularly distinctive character.[66] Stripped of the formal support of tonal conventions, the unfolding panoply of heightened expressive moments gives a figure to the disorienting sense of psychic dislocation that at one time authorized psychoanalytic or quasi-autobiographical readings.[67] A succession of explosive moments evokes the traumatic shock of a seismic breach in the individual's and history's presumed self-same identity. The resulting temporal fissures thus punctuate the feeling of welling anxiety programmatically ascribed to the Woman's psychic desolation.

When the quotation from Schoenberg's *Am Wegrand* appears at the climactic moment in inverted form, the distance separating *Erwartung*'s disconsolate world from the familiar terrain of tradition tonal works is perhaps most acute.[68] Here, when the search for her lover approaches the monodrama's denouement, the Woman's declamation "*Tausend Menschen ziehen vorüber ... ich kenne dich nicht*" (Example 4.2) anticipates the moment of recognition when she discovers her dead lover's body. In this moment of recognition, the world of *Erwartung*'s psychodrama crystallizes with the work's singular revelation of the modern individual's despair. It is as though, exiled from an ontotheological order, the self, history, and even thought about time are thrown into crisis by the loss of a confidence in the modernist ideals of progress and the promise of better times to come.

Adorno's criticism of the second-rate Expressionism from which *Erwartung* suffers, for him, due to its connection with Pappenheim's text, reveals more about Adorno's critical intransigence than it does about *Erwartung* in this regard. By turning the work's figuration of dissonance into the "technical law of music's form,"[69] Adorno effectively subordinates the work's expressive features to modern music's negative dialectical *raison d'être*. In Adorno's view, Schoenberg forces his music to repeat its expressive gestures through drawing its inner form from Pappenheim's text and by adapting his music to it. By violating the "postulate of incessant innovation,"[70] *Erwartung* thus ostensibly betrays its own ideal. Adorno admits that sustaining the shock of the new at every instant presents an impossible task so that the "utopia of the pure τόδε τι [a separate individuated substance]"[71] is constantly beyond reach. At the same time, the formal technical law that he champions becomes hostage to the only philosophy that he believed could be practiced responsibly in the face of despair in order to contemplate "all things as they would present themselves from the standpoint of redemption."[72]

As the hibernatory refuge of truth, music's and art's emphatic negation of meaning consequently remonstrates against the existing order of material life without for all that being able to offer any alternative. This hibernatory refuge is the vanishing point of modern music's and art's claim to truth—hence, their melancholic condition. According to Adorno, the liquidation of all stylistic conventions reduces expression to the immediate locus of its own absolute impulse. Yet works speak only by gathering these impulses together. The notion that "[t]otal contradiction is nothing but the manifested untruth of total identification"[73] foregoes the fact that a work's subversion

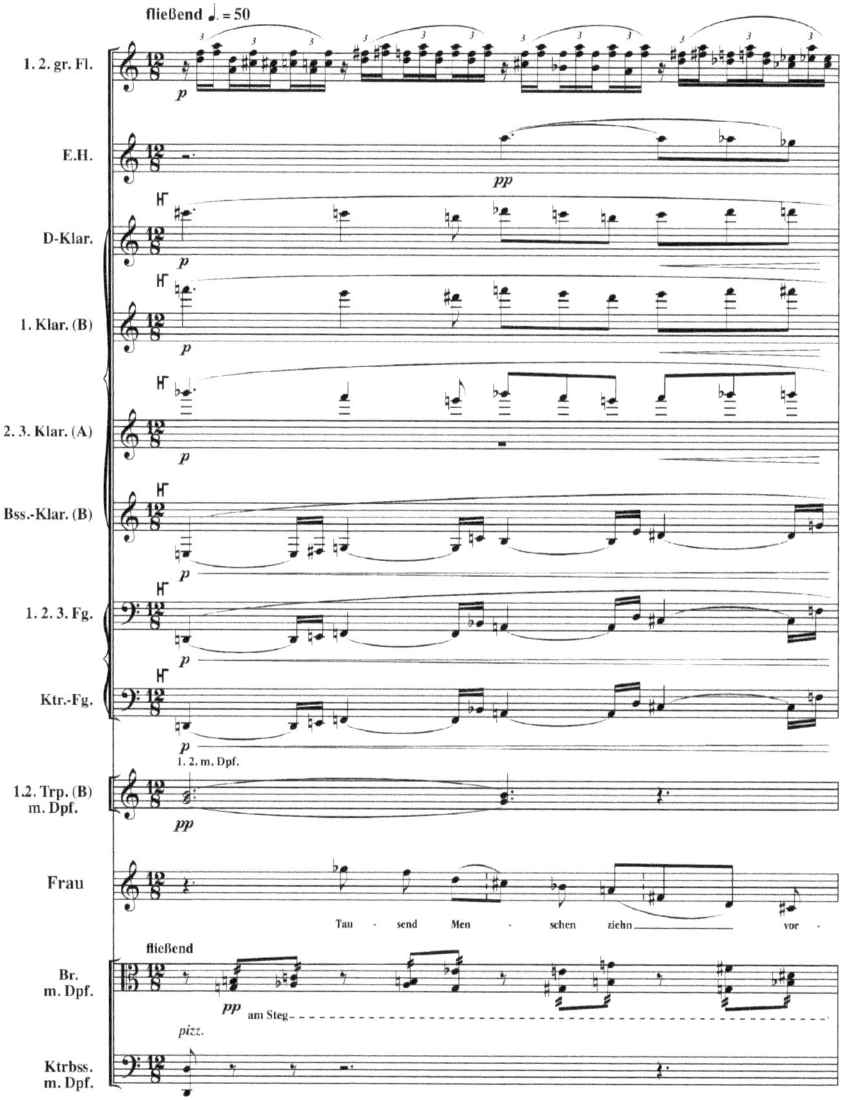

Example 4.2 Arnold Schoenberg, *Erwartung*, mm. 411–413.

Source*: Arnold Schönberg "Erwartung|Monodram in 1 Akt|für Sopran und Orchester|op. 17"*

Example 4.2 (Continued).

of ideologically congealed systems and habits of thought is one that the work effects by virtue of its worlding power. Modern music's and art's melancholic condition is therefore in the end symptomatic of the performative contradiction that inheres in the radical negation of a style of philosophizing that dares to master the meaning of every present through grasping time in its totality in an eternal present. Contrariwise, in answer to the negation of tonality's regulative order, and in contradistinction to Adorno's assessment, *Erwartung*'s expression of anxious foreboding has the force of an event comparable to the one brought about by the Hegelian philosophy of history's loss of credibility.

Negative ontotheology, the "bad infinity," and the logic of hope

By likening *Erwartung*'s compositional *tour de force* to the event in thinking precipitated by a critique of the philosophical act of faith that vests the truth of its system with its rational authority, I want to highlight how Schoenberg's renunciation of tonal closure brings the force of the present sharply into focus. An emblem of the crisis born from the negation of tonality's unifying force, *Erwartung*'s evocation of anxious foreboding places the modernist assurance of the present's continuity with the past and the future radically into doubt. The negation of tonality's normative function radically subverts the tonal system's emblematic, ontotheological value. At the same time, the aporia unleashed by emancipating the dissonance from the requirement that it resolve fuels the demand for expression that ignites *Erwartung*'s explosive response. Tonality's negation by analogy does not therefore necessary amount to replacing a totalizing system of thought with its negative inverse. On the contrary, Schoenberg's renunciation of a tonal center stands out as a crucial event not least because it marks a critical distinction between the cultural and philosophical conceits of such a system of thought and the negative inverse of this system of thought's despair of the logic of hope.

The philosophical event announced by denunciations of the Hegelian philosophy of history sets thought about the history's temporalization against a system that dares to totalize time in the eternal present. By measuring itself against the past's, the present's, and the future's speculative unity, this event countermands the philosophical act of faith that reason is an infinite force that produces the historical circumstances necessary for its own realization. Ricoeur reminds us that the Hegelian temptation rests on a style of thinking that dares to "elevate itself to grasping history as the totalization of time in the eternal present"[74] by embracing the past, the present, and the future as a singular whole. For the speculative philosopher, this act of faith's authority is consubstantial with the authority of the system's self-presentation (*Selbstdarstellung*). The cunning with which reason guides the historical actors' intentions without their knowing it is the "apologetic doublet"[75] of this philosophical credo. Hence, the Hegelian thesis regarding the cunning of reason takes the place that "theodicy assigns to evil when it protests that evil is not in vain."[76]

Equating freedom with Spirit's fulfillment of its rationally necessary plan paradoxically places the temporalization of history under the rule of the eternal present. Ricoeur stresses that a critique that is worthy of the challenge posed by Hegel must confront the idea that thought can rise to the level of the absolute by grasping the "underlying unity of the surpassed past and the coming manifestations of life"[77] that announce themselves in signs that we already understand. By setting history's temporalization against the eternal present of Spirit, Ricoeur ties the force of the present to the imperfect mediation between a past that has already been surpassed and the horizon of our expectations (as I outlined earlier). Ricoeur therefore reminds us of the ethical and political obligation to preserve the tension between the space of our experiences and the horizon of our expectations.[78]

The logic of hope accentuates how the force of the present temporalizes the time of history in countermanding the Hegelian temptation. Expectations fueled by unfulfilled demands animate the hope that the present holds as yet unrealized possibilities in readiness. Ricoeur's post-Heideggerian return to Kant through Hegel can therefore be credited at least in part to the way that hope's practical and existential necessity inscribes itself in the structure of action. Ricoeur explains that the concept of freedom's actualization has two dimensions: one of realized accomplishments and another of as yet unfulfilled demands. Hegel's system is "one that is written from the end toward the beginning,"[79] such that its rationality coincides with the whole system's recapitulation in an eternal present. Kant's philosophy links reason's demand for totality to the impossibility and even the prohibition of positing any such totality in advance. Hence, a "horizon of unfulfilled claim belongs [as much] to the most genuine experience of action"[80] as do accomplished achievements. The hope of as yet unfulfilled expectations consequently ignites the passion for the possible. In turn, this passion for the possible is hope's answer to all *amor fati*. For, in response to "all Nietzschean love of destiny, [and] to all worship of fate,"[81] hope summons the "courage to be."[82]

In view of the preceding remarks, it might be tempting to regard *Erwartung*'s expression of its solitary figure's fragmented reality as the twentieth century's musical exemplar of disconsolate despair. By setting *Erwartung*'s temporally enigmatic character against the triumphant order of Beethoven's *Eroica*, Berthold Hoeckner underscores this twentieth-century work's negative inverse relation to time's totalization within the Hegelian system of thought. According to Hoeckner, *Erwartung*'s temporal paradox inverts that of the *Eroica* through converting into its opposite the "heroic fusion of an expansive reality into the eternal moment of [the Hegelian system's] 'good infinity.'"[83] For Hoeckner, the dream of mastering time via its totalization in an eternal present is thus transformed into a nightmare of the "bad infinity" of a frozen instant sempiternally suspended above the abyss. It is as though by evoking the "eternity of a single instant"[84] over its unfolding course, *Erwartung* becomes the negative riposte to the totalization of time's fraudulent onto-theological promise.

The "bad infinity" of this frozen instant sets the force of the present starkly in relief. In response to the Hegelian challenge with respect to thinking about the oneness of time, the present that acquires its force as the time of our initiative revokes the "bad infinity" of a sempiternal present suspended between past and future horizons. By the same token, the impossibility of totalizing time within the sphere of our historicality shatters the "good infinity" that Hoeckner identifies with the Hegelian motif of Beethoven's music. It is as though *Erwartung* countermands the Hegelian temptation through intensifying the aporia that springs from the dissociation of the three temporal ecstasies (future, past, and present). This aporia shipwrecks the hubris of a mode of thinking that dares to master thought about time by grasping history as a whole in the eternal present. *Erwartung* no doubt embodies the determinate negation of the dream of a reconciled humanity through crystallizing in a single moment of despair. Yet we might also wonder if in ushering in the "dusk of modernity,"[85] *Erwartung* wrests the dream of a reconciled humanity from the temptation either to posit this dream's fulfillment in advance of its historical realization or to abandon it altogether.

Should we conclude that the "bad infinity" of an eternal instant cast into the abyss of time marks the descent into a negative ontotheology? *Erwartung*'s apparent reversal of the Hegelian ideal seems to amount to the radical negation of tonality's ontotheological thrust. Yet *Erwartung*'s expression of foreboding anxiety (which is redoubled by the frustrated search for some hidden tonal order[86]) equally stands as witness to the crisis of a present torn between two fleeing horizons (that of the past and of the future), which *Erwartung* places on stage. By giving heightened expression to the experience of time as a principle of the self's "alienation and dispersion,"[87] *Erwartung* sharpens the contrast between time and eternity. Hence, its figuration of the eternity of the second is more than just the negative inverse of the "good infinity." It is as if *Erwartung*'s personification of the experience of time as the principle of the self's dispersion and alienation in the figure of the solitary Woman, who is incapable of constructing a temporally coherent identity from her disordered psychic states, also embodies the moment of crisis when the present, suspended between a past that has already been surpassed and a future that withdraws more quickly than it can be approached, is wholly torn within itself.

Might *Erwartung*'s personification of a time of crisis also sound a note of hope? Ernst Bloch for one hears a clarion utopian call issuing from the recesses of Schoenberg's dissonant music. The triumphant *tuba mirum spargens sonum* that in Beethoven's *Fidelio* heralds Florestan's deliverance here gives way to the long anticipations and suspended arrivals that reflect the "hopelessness of the bourgeoisie and … its interest in sapping all will for change."[88] Schoenberg's music is accordingly full of the "still undefined spark-figure"[89] of utopian longing fomenting in the hollow space of the present. Bereft of any tonal point of reference or tonic homeland, his music clears a space for the chiliastic leap marshalled by the forces of Marxist science that led Bloch to embrace this science as the "quartermaster of the future."[90] In contrast to Adorno's interpretation of Schoenberg's music, which redoubles the virulence of his negative dialectical critique, Bloch's reading interposes his principle of hope. For Bloch, the boldness with which Schoenberg's works venture to do justice to the ideas that they express thus sets these works' dissonant worlds against the nihilism of hope's despair. Yet this principle of hope's Marxist millennialist thrust holds out the promise of a future at the cost of any historical handholds.

Perhaps the nadir of hope's despair could be represented fictionally only in a novel where the crisis personified by *Erwartung* ends in a Faustian pact. By following the "dire consequences of the constructive Schönbergian approach to music"[91] to its conclusion, Thomas Mann in his novel *Doctor Faustus* depicts the dark, mythological realm into which Mann believed art had been cast back. For Mann, music as treated in the novel provides a vehicle for expressing the general situation of art and culture at the time. Schoenberg's twelve-tone method served as a model of an art from which "life has vanished."[92] The *Lamentations of Dr. Faustus* by Adrian Leverkühn—the novel's protagonist—is the emblem of the novel's central motif: "the closeness of sterility, [and] the innate despair that prepares the ground for a pact with the devil"[93] spawned by the cultural crisis in which music was also caught up. Leverkühn's "Ode to Sorrow" consequently negates Beethoven's exultant "Ode to Joy" through surrendering to the fateful dialectic of mimesis and construction.[94] The demonic jest that liberates music's expression by means of the most severe of calculating constructions echoes the subject's mimetic identification with the anthropogenic forces that annihilate it. The echo effect that in Leverkühn's *Lamentations* constitutes a principle of construction renders fugue meaningless; the monstrous inverse of fugue's aesthetic prefiguration of the fullness of the present *hic et nunc*, Leverkühn's *Lamentations* is the emblem of the negative sublime.[95] Since every note is subject to the most rigorous calculation, "every transformation is itself [only] an echo of what preceded it."[96] This principle of construction's anthropogenic substance thus redoubles music's mythic origin as the expression of lament, but only as the demonic counterpart of the *nunc stans*.

Following Adorno's advice, Mann did not leave Leverkühn's *Lamentations* unfinished. For up to the very end, there could be "no consolation, reconciliation, [or] transfiguration."[97] It may be that Leverkühn's "negation of the religious"[98] in his *Lamentations* stands witness to sempiternal suffering. However, Serenus Zeitblom (the novel's narrator) explains that Leverkühn's last work does not deny the religious: a "work dealing with the Tempter, with apostasy, with damnation—how can it be anything but a religious work!"[99] But how, we could ask, does this tortured dialectic of the demonic and the divine counter Nietzsche's hedonistic nihilism with Adorno's apocalyptic aesthetic? The final G in the high violin, which we hear only through the words of the novelist, echoes in the silence into which the recollection of the hope for reconciliation falls.[100] In dying and disappearing, this final tone reveals the absence of the absolute. Like negative dialectics' solidarity with "metaphysics at the time of its fall,"[101] the metaphysics of German music at this music's end here succumbs to the crisis that follows in the wake of the catastrophic betrayal of the Enlightenment's emancipatory project.

This betrayal of the dream of a reconciled humanity, however, is not the final word. Deprived of the consolation of traditional paradigms, we cannot avoid becoming aware of the violence that ensues when the social or political imposition of a heteronomous order replaces the dissonance and discord that springs from the contestation of existing systems and mores. Yet, if by "succumbing to the fascination of the absolutely unformed and to the plea for that radical intellectual honesty Nietzsche called *Redlichkeit*"[102] we forfeit the capacity to preserve the tension between the space of present experiences and the horizons of our expectations, do we not also surrender our power to take an initiative in response to the demands of the situations in which we find ourselves? The radical inversion of the *nunc stans* in Leverkühn's *Lamentations* draws near the tragedy of the sempiternal, which scatters the force of the present in the void between the past and the future. At the same time, however, even this negative counterimage of the *nunc stans* preserves the ineliminable trace of the enigma of the being and the nonbeing of time. The "bad infinity" of the eternity of the instant does not in the end abolish time. On the contrary, the crisis announced through arresting the present in the eternity of this frozen instant gives rise once again to a reflection on the force that the present has in answer to time's distension and dispersion.

In Chapter 6, I will explore how representations of the "end of time" simulate the "standing present" in response to the enigma of time's ultimate unrepresentability. Schoenberg's negation of tonality's nomic principles marks out a limit-situation in this respect. In contrast to the presentiment of abiding grace and in opposition to a mode of thought that dares to master time, his music gives voice to the crisis of a present suspended between two fleeing horizons. Leverkühn's lament, "Ode to Sorrow," amplifies the force of this crisis as a fictive testament to the grief that sets all hope at a seemingly infinite distance. In response to the affliction of suffering and despair, lament, I will argue in the next chapter, initiates the work of mourning.

Notes

1 Arnold Schoenberg, *Style and Idea: Selected Writings of Arnold Schoenberg*, ed. Leonard Stein, trans. Leo Black (Berkeley, CA: University of California Press, 1975), 86.
2 Schoenberg famously recounts that when asked by his superior officer whether he was the "notorious Schoenberg," he replied: "Beg to report, sir, yes.... Nobody wanted to be, someone had to be, so I let it be me" (Schoenberg, *Style and Idea*, 104).
3 Schoenberg, *Style and Idea*, 86.
4 Leo Treitler, *Music and the Historical Imagination* (Cambridge, MA: Harvard University Press, 1989), 97 ff. This model appeals to the explanatory power of the natural sciences. Deducing the occurrence of an event from established premises thus has the logical value of a positivist explanation. See Paul Ricoeur, *Time and Narrative*, vol. 1, trans. Kathleen McLaughlin and David Pellauer (Chicago, IL: University of Chicago Press, 1984). Under the covering law model, "the notion of a historical event was divested of its narrative status and placed within the framework of an opposition between particular and universal" (112) so that historical events were placed under a general concept that included physical events and occurrences. The first premise of the covering law model accordingly "describes the initial conditions: prior events, prevailing conditions, and the like. The second states a regularity of a certain type, that is, a hypothesis of a universal form, that, if verified, merits being called a law" (112).
5 Carl Dahlhaus, *Foundations of Music History*, trans. J. Bradford Robinson (New York: Cambridge University Press, 1983), 19–20; see 28 ff.
6 Ethan Haimo, *Schoenberg's Transformation of Musical Language* (New York: Cambridge University Press, 2006), 353; see Ernst Bloch, *Principle of Hope*, vol. 3, trans. Neville Plaice et al. (Oxford: Basil Blackwell, 1986), 1071; Luigi Rognoni, *The Second Vienna School: Expression and Dodecaphony* (*The Rise of Expressionism in the Music of Arnold Schoenberg, Alban Berg, and Anton von Webern*), trans. R. W. Mann (London: John Calder, 1977).
7 See my *Structure and Sorcery: The Aesthetics of Post-War Serial Composition and Indeterminacy* (New York: Garland, 1989). Boulez's radicalization of Schoenberg's replacement of the tonal system with the twelve-tone method in *Structures 1a* underscores the hyper-rational consequences of reducing the musical "discourse" to its zero point. See also Treitler, *Music and the Historical Imagination*. Treitler points out that Western art music's supposedly arrested evolution spawned two major developments. The French solution, represented by the "expansion of the concept of consonance" (124) by Gabriel Fauré, Claude Debussy, and Maurice Ravel, stabilized individual sonorities, producing new harmonies from new tonal combinations such as eleventh chords. The central European solution, which stemmed from Arnold Schoenberg's "abandonment of the consonance-dissonance concept" led to replacing the principle of tonal coherence with that of the twelve-tone method. This method ostensibly replaced the structural significance of the "*relations* between tones" with that of their "*succession*" (original emphasis). On this view, total serialism is the logical successor to Schoenberg's emancipation of dissonance. Serialism was superseded in turn by the introduction of chance techniques and by developments in *musique concrète* and electronic music.
8 Joseph Kerman, "How We Got into Analysis, and How to Get Out," Critical Inquiry 7, no. 2 (1980), 318. Kerman emphasizes that the

> ideology [of organicism] did not receive its full articulation until the music in which it was rooted came under serious attack. This occurred around 1900

when tonality, the seeming linchpin of the entire system, began to slip in Germany as elsewhere.

(316)

9 Carl Dahlhaus, *Schoenberg and the New Music*, trans. Derrick Puffett and Alfred Clayton (New York: Cambridge University Press, 1987), 14.
10 Jürgen Habermas, *The Philosophical Discourse of Modernity: Twelve Lectures*, trans. Frederick Lawrence (Cambridge, MA: MIT Press, 1987), 32. Habermas attributes this philosophical self-understanding to Friedrich Hölderlin and Friedrich Wilhelm Joseph Schelling as well as to Hegel.
11 Charles Taylor, *Hegel* (Cambridge: Cambridge University Press, 1975), 469. See note 38, Chapter 1.
12 Kerman, "How We Got into Analysis," 315.
13 Georg Wilhelm Friedrich Hegel, *Lectures on the Philosophy of World History. Introduction: Reason in History*, trans. Hugh Bar Nisbet (Cambridge: Cambridge University Press, 1975), 52.
14 Heinrich Schenker, *Free Composition (Der freie Satz): Volume III of New Musical Theories and Fantasies*, trans. Ernst Oster (New York: Longman, 1979), 3. For Schenker, the union of foreground, middle-ground and background accordingly "expresses the oneness of an individual, self-contained life." By reducing fore- and middle-ground linear progressions to their essential background structure, Schenker believed he had uncovered the principle of the preternatural processes at work in the compositions of the great masters. At the deepest level, the harmonic progression outlined by the movement from the tonic (I) through the dominant (V) and returning to the tonic typically supports a structurally prolonged descent of the third scale degree through the second to the first. This fundamental *Ursatz* ostensibly guarantees that the surface events unfold in a manner consistent with the expression of the work's inner, organic life. Schenker likened a work's tonal development to the life history of a civilization. Just as in a tonal work the "*goal* and the course to the goal are primary" (5), a community's or civilization's history manifests its native predisposition according to the inner laws from which its development naturally flows.
15 Schenker, *Free Composition*, xxiv. For Schenker, the "power of will and imagination which lives through the transformation of a masterwork reaches us in our spirit as a power of imagination—whether we have specific knowledge of the fundamental structure and the transformations or not" (6). His intent was to give "the fullest possible expression to what the music of the masters was and must continue to be if we wish to keep it alive" (xxii). Schenker's approach lays the ground for the ideal of "structural listening" that Theodor Adorno places at the center of his typology of listeners. "Structural listening" consequently formalizes the "product" of the configurating operation on which the listening experience depends. See Theodor W. Adorno, "On the Problem of Musical Analysis," trans. Max Paddison, *Music Analysis* 1, no. 2 (1982); Theodor W. Adorno, *Introduction to the Sociology of Music*, trans. E. B. Ashton (New York: Continuum, 1989); Rose Rosengard Subotnik, *Deconstructive Variations: Music and Reason in Western Society* (Minneapolis, MN: University of Minnesota Press, 1996).
16 Paul Ricoeur, *Time and Narrative*, vol. 3, trans. Kathleen Blamey and David Pellauer (Chicago, IL: University of Chicago Press, 1988), 194.
17 See my discussion of music's metaphysical role as the refuge of a bourgeois *mentalité* in the Introduction.
18 Susan McClary, *Feminine Endings: Music, Gender, and Sexuality* (Minneapolis, MN: University of Minnesota Press, 1991), 107; see Susan McClary, "Narrative

Agendas in 'Absolute Music': Identity and Difference in Brahms's Third Symphony," in *Musicology and Difference: Gender and Sexuality in Musical Scholarship*, ed. Ruth A. Solie (Berkeley, CA: University of California Press, 1993); Susan McClary, "Paradigm Dissonances: Music Theory, Cultural Studies, Feminist Criticism," *Perspectives of New Music* 32, no. 1 (1994).

19 McClary, *Feminine Endings*. For McClary, tonal closure is a cultural guarantor of the "masculine" protagonist's identity, and, by framing the contagion of the Other, it also makes the Other an object of surveillance and fascination (15; 80 ff.).

20 Georg Wilhelm Friedrich Hegel, *Hegel's Phenomenology of Spirit*, trans. A. V. Miller (Oxford: Oxford University Press, 1977), 11.

21 Theodor W. Adorno, *Minima Moralia: Reflections from Damaged Life*, trans. Edmund F. N. Jephcott (London: Verso, 1974), 50. Adorno here inverts Hegel's thesis.

22 Hegel, *Phenomenology of Spirit*, 11:

> Of the Absolute it must be said that it is essentially a *result*, that only in the *end* is it what it truly is; and that precisely in this consists its nature, viz. to be actual, subject, the spontaneous becoming of itself.
>
> (original emphases; see 263 ff.)

The coherence of this system underscores the seemingly rational force vested in tonal closure. Hence, the conceit: A tonal work that by analogy confers the idea of freedom's actualization on the totality of its movement repeats the metaphysical claim of the subject to posit itself as master of meaning through elevating thought about time to the level of the absolute.

23 Schoenberg, *Style and Idea*:

> Bach, who knew more secrets than the Netherlanders ever possessed, enlarged these rules to such an extent that they comprised all the twelve tones of the chromatic scale. Bach sometimes operated with the twelve tones in such a manner that one would be inclined to call him the first twelve-tone composer.
>
> (117)

24 Carl E. Schorske, *Fin-de-Siècle Vienna: Politics and Culture* (New York: Vintage Books, 1981), 345. Schorske points out that Schoenberg shared a "diffuse sense that all is flux, that the boundary between ego and the world is permeable" with his older contemporaries, Hugo von Hofmannsthal, Sigmund Freud, Gustav Klimt, and Ernst Mach, who were intellectual pioneers of Vienna's elite.

25 Hans-Georg Gadamer, *Praise of Theory: Speeches and Essays*, trans. Chris Dawson (New Haven, CT: Yale University Press, 1998), 27.

26 Gadamer, *Praise of Theory*, 27.

27 Ibid.

> The scientifically shaped cultural consciousness of modernity has certainly gone hand in hand with an increasingly active critique of culture, which intensified still further at the beginning of our century. As alienation from modern labor and cultural life (Bildung) grew, various protest movements appeared, such as the great romantic children's crusade of the youth movement before the First World War, which already marked the threshold of the technological age. The full outbreak of this age was the technological slaughter of the First World War.... The eight decades we can survey of our century's world wars and world crises seem strangely unified under this universal aspect, presumably because philosophical consciousness has

> remained fairly constant since the crisis of the First World War. The consciousness of the time that it reflects has, in the end, persisted to an astonishing extent through all the to and fro of short-term variations. The First World War shattered Europe's self-consciousness.... The endangered self-consciousness of our epoch is no longer defined by the shifts in these [America and Russia's] powers' political interrelations or the fundamental dissimilarity of their economic and social systems, as by the inexorable law of progressive industrialization, which applies in the same way to the both of them.
>
> (27–28)

28 Hannah Arendt, *Between Past and Future: Eight Exercises in Political Thought* (New York: Penguin Books, 1968), 13; see Hannah Arendt, *The Promise of Politics*, ed. Jerome Kohn (New York: Schocken Books, 2005); Hannah Arendt, *The Human Condition* (Chicago, IL: Chicago University Press, 1958).
29 See Ricoeur, *Time and Narrative*, vol. 3, 207.
30 Arnold Schoenberg, *Theory of Harmony: 100th Anniversary Edition*, trans. Roy E. Carter, forward by Walter Frisch (Berkeley, CA: University of California Press, 1978), 49.
31 Theodor W. Adorno, *Aesthetic Theory*, ed. Gretel Adorno and Rolf Tiedemann, trans. and ed. Robert Hullot-Kentor (Minneapolis, MN: University of Minnesota Press, 1997). For Adorno: "Suffering remains foreign to knowledge; though knowledge can subordinate it conceptually and provide means for its amelioration, knowledge can scarcely express it through its own means of experience without itself becoming irrational" (18; see 110).
32 Bryan Hyer, "Tonality," in *The New Grove in Dictionary of Music and Musicians*, vol. 25 (Taiwan to Twelve Apostles), ed. Stanley Sadie (London: Macmillan Publishers Limited, 2001).

> For [François-Joseph] Fétis, the dominant 7th was the crucial musical element in *tonalité moderne*, the "birth" of which he registered in a Monteverdi madrigal, *Stracciami pur il core*, from Book 3 of 1592 (*Equisse de l'histoire de l'harmonie*, 1840).... [Fétis] later made the same claim ... about another madrigal, *Cruda Amarilli*.... Both [Alexandre] Choron and Fétis ... cite the birth of the dominant 7th in the music of Monteverdi as the decisive event in the historical separation of *tonalité moderne* from *tonalité antique* (Choron) or *ancienne* (Fétis).
>
> (584; 592)

See Gary Tomlinson, *Monteverdi and the End of the Renaissance* (Berkeley, CA: University of California Press, 1987). Tomlinson points out that the last six pieces in Monteverdi's Fifth Book of madrigals "are true continuo madrigals. In writing them Monteverdi began to explore the novel expressive and structural capabilities of the new medium" (152).
33 Dahlhaus, *Schoenberg*, 77.
34 Ibid.
35 Schoenberg, *Style and Idea*, 216.
36 Ibid., 91.
37 Ibid.
38 Schoenberg, *Style and Idea*, 91; see 216 ff. Schoenberg consequently claims that a musical poet's—a musical thinker's—representation of musical ideas in her or his works "must correspond to the laws of human logic ... [since] they are a part of what man can apperceive, reason and express" (220). For him, the

> conviction that order, logic, comprehensibility and form cannot be present without obedience to these laws... forces the composer along the road of exploration. He must find, if not laws or rules, at least ways to justify the dissonant character of these harmonies and their successions.
>
> (218)

The twelve-tone method's aesthetic and theoretical necessity thus elevated this method above a "mere technical device to the rank and importance of a scientific theory" (220).

39 Charles Rosen, *Arnold Schoenberg* (New York: The Viking Press, 1975), 26–27.

40 Rosen, *Schoenberg*, 44.

41 Schoenberg, *Style and Idea*, 123; original in italics.

42 Schoenberg, *Style and Idea*, 123. Schoenberg accordingly identifies the principle of developing variations with the

> variation of the features of a basic unit [that] produces all the thematic formulations which provide for fluency, contrasts, variety, logic and unity, on the one hand, and character, mood, expression, and every needed differentiation, on the other hand—thus elaborating the *idea* of the piece.
>
> (397; original emphasis)

See Dahlhaus, *Schoenberg*, 133.

43 Arnold Schoenberg, quoted in Patricia Carpenter and Severine Neff, "Commentary," in *The Musical Idea and the Logic, Technique, and Art of Its Presentation*, trans. and ed. Patricia Carpenter and Severine Neff (Bloomington, IN: Indiana University Press, 2006), 15; original emphasis. The original passage is included in Appendix 3: "German Texts of Unpublished *Gedanke* Manuscripts Referred to in the Commentary."

44 Schoenberg, quoted in Carpenter and Neff, "Commentary," 15; original emphasis. See Dahlhaus, *Schoenberg*. For Dahlhaus, the notion that the "intelligibility of the individual feature depends on the logic of the whole" (77) is key to Schoenberg's musical poetics. Hence, for Dahlhaus, the idea that "assumes concrete form in a work... is realized less in the musical shapes that make up the surface than in the tissue of relationships which, hidden beneath, connect the ideas with one another" (81).

45 Carpenter and Neff, "Commentary," 3. Carpenter and Neff emphasize that Schoenberg's conception of the musical idea aimed as much at differentiating between his ideas of tonal and twelve-tone composition as it served to ensure their continuity. Notwithstanding their assertion that the musical idea cannot be abstracted from a work, Schoenberg's theoretical proclivities led him to distinguish the musical idea's many-sided presentations (*Gestalten*) from these presentations' plenipotentiary source (*Grundgestalt*).

46 Carpenter and Neff, "Commentary," 3.

47 Schoenberg, *Style and Idea*, 123.

48 Ibid.

49 Theodor W. Adorno, *Prisms*, trans. Samuel and Shierry Weber (Cambridge, MA: MIT Press, 1967), 154.

50 Theodor W. Adorno, *Negative Dialectics*, trans. E. B. Ashton (London: Routledge and Kegan Paul, 1973), 5. Hence: "Contradiction is nonidentity under the aspect of identity" (5).

51 Adorno, *Negative Dialectics*, 6.

52 Adorno, *Aesthetic Theory*, 110. "The emancipation from the concept of harmony has revealed itself to be a revolt against semblance: Construction inheres tautologically in expression, which is its polar opposite" (100).

53 Theodor W. Adorno, "Commitment," in *Aesthetics and Politics: The Key Texts of the Classic Debate Within German Marxism* [Collected Essays by

Theodor Adorno, Walter Benjamin, Ernst Bloch, Bertolt Brecht, George Lucas], (London: Verso, 1977), 194.

54 Dahlhaus, *Schoenberg*, 87.

55 See Chapter 7.

56 See Bryan R. Simms, "Whose Idea Was *Erwartung*?" in *Constructive Dissonance: Arnold Schoenberg and the Transformations of Twentieth-Century Culture*, ed. Juliane Brand and Christopher Hailey (Berkeley, CA: University of California Press, 1997). Simms comments that "Eva Weissweiler has argued persuasively that Pappenheim's *Erwartung* is essentially an attack upon the Viennese upper class, of which she sees the Woman as representative" (103). He suggests that it

> is very unlikely that Schoenberg interpreted the text of *Erwartung* as Pappenheim had intended, as a realist study of hysteria with feminist overtones. Instead his own reading of the poem focused directly on the jumble of emotions that the Woman experiences.
>
> (103)

57 Schorske, *Fin-de-Siècle Vienna*, 354.

58 Ibid.

59 Joseph Auner, "'Heart and Brain in Music': The Genesis of Schoenberg's *Die glückliche Hand*," in *Constructive Dissonance: Arnold Schoenberg and the Transformations of Twentieth-Century Culture*, eds. Juliane Brand and Christopher Hailey (Berkeley, CA: University of California Press, 1997), 118. Auner comments that "at the time of *Erwartung*, his [Schoenberg's] reliance on intuition was accompanied by a deep ambivalence about the value of order and logic in composition." See Joseph Auner, "Schoenberg's Aesthetic Transformation and the Evolution of Form in *Die glückliche Hand*," *Journal of the Arnold Schoenberg Institute* 12, no. 2 (1989). Schoenberg's remark in a letter to Kandinsky seems to support this interpretation, when he writes that art "belongs to the *unconscious*!" (Schoenberg, quoted in Holly Watkins from a letter from Schoenberg to Wassily Kandinsky, 1911 in "Schoenberg's Interior Designs," *Journal of the American Musicological Society* 61, no. 1 [2008]: 135). See Rognoni, *The Second Vienna School*, 13. See also Simms, "Whose Idea was *Erwartung*?" Simms quotes Schoenberg's statement that the composer could "no more control the musical outcome than the hysterical patient can control the symptoms of her ailment" (quoted in Simms, "Whose Idea Was *Erwartung*?," 106). According to Simms, "Schoenberg's theory of emotions is … a key to understanding the highly experimental musical language of *Erwartung*, which deviates so strikingly from his music prior to 1909" (106). Schoenberg's focus on emotions in *Erwartung* is accordingly "symptomatic of his broader interest in the human psyche as the source of artistic creativity and his own atonal musical language" (105).

60 See my discussion of the language of music in my *Hermeneutics and Music Criticism* (New York: Routledge, 2010), 86 ff.

61 Simms, "Whose Idea Was *Erwartung*?," 105.

62 Peter Bürger, *Theory of the Avant-Garde*, trans. Michael Shaw (Minneapolis, MN: University of Minnesota Press, 1984), 56; see 33–34. The avant-gardiste work negates "a specific kind of unity, the relationship between part and whole that characterizes the organic work of art" (56). The avant-garde's self-criticism legitimated its intended negation of art as an institution separate from social *praxis*. While the social subsystem founded on Aestheticism's claim of aesthetic autonomy fulfilled a positive role by severing art's connection with society, the separation of art and life codified by Valéry's principle of *l'art pour l'art* deprived the artist of any social function. Modernist avant-garde movements intended to combat art's isolation by capitalizing on the positive side of the process whereby

the logic of bourgeois society's development, and the increasing division of labor and social specialization, instituted the break by virtue of which aesthetic experience conflicted with this developmental logic's means-ends rationality.

63 Bürger, *Theory of the Avant-Garde*, 82.

64 Ibid.

65 Bürger, *Theory of the Avant-Garde*, 82: "For the act of reception, this means that even the avant-gardiste work is still to be understood hermeneutically (as a total meaning) except that the unity has integrated the contradiction within itself."

66 See Hans-Georg Gadamer, *The Relevance of the Beautiful and Other Essays*, ed. Robert Bernasconi, trans. Nicholas Walker (Cambridge: Cambridge University Press, 1986). Gadamer emphasizes that we "cannot recuperate within the concept the meaningful content that addresses us in art" (33). See also Rosen, *Schoenberg*. Rosen comments that *Erwartung* has traditionally been the "despair of musical analysis" (40). Martha M. Hyde similarly remarks upon the significant difficulty that Schoenberg's "contextual" works present for analysis. Martha M. Hyde, "Musical Form and the Development of Schoenberg's Twelve-Tone Method," *Journal of Music Theory* 29, no. 1 (1985).

67 The connection between *Erwartung*'s portrayal of the Woman's increasing anxiety and Freudian psychoanalysis has largely been discredited. See for example Watkins, "Schoenberg's Interior Designs"; Lewis Wickes, "Schoenberg, *Erwartung* and the Reception of Psychoanalysis in Musical Circles in Vienna until 1910/11," *Studies in Music* 23 (1989); Alexander Carpenter, "Schoenberg's Vienna, Freud's Vienna: Re-Examining the Connections between the Monodrama *Erwartung* and the Early History of Psychoanalysis," *Musical Quarterly* 93, no. 1 (2010). See also Schorske, *Fin-de-Siècle Vienna*. According to Schorske, Schoenberg projected into *Erwartung* the shattering personal experience of his wife's abandonment "in favor of one of his best friends, the artist Richard Gerstl, who shortly thereafter committed suicide" (354). Schorske acknowledges, however, that such personal anguish cannot explain artistic creation.

68 See Berthold Hoeckner, *Programming the Absolute: Nineteenth-Century German Music and the Hermeneutics of the Moment* (Princeton, NJ: Princeton University Press, 2002), 205 ff. For Hoeckner, Herbert Buchanan's and Alan Lessem's

> search for tonal order [in reference to the *Am Wegrand* quotation] betrays an anxiety similar to that of the woman's search for her lover in in *Erwartung*.... Tonality thus surfaces like a repressed memory of the hidden desire that the unsung words from *Am Wegrand* had expressed as a "longing" which "fulfillment will not fulfill."
>
> (208)

See also Herbert H. Buchanan, "A Key to Schoenberg's *Erwartung* (Op. 17)," *The Journal of the American Musicological Society* 20, no. 3 (1967); Alan Philip Lessem, *Music and Text in the Works of Arnold Schoenberg: The Critical Years, 1908–1922* (Ann Arbor, MI: UMI Research Press, 1979). Lessem underscores the close resemblance between *Erwartung*'s motivic materials and those of Schoenberg's setting of John Henry Mackay's poem in the song *Am Wegrand* (1905) and which was later included in his *Eight Songs*, op. 6. See also Bryan R. Simms, *The Atonal Work of Arnold Schoenberg, 1908–1923* (New York: Oxford University Press, 2000), especially 97 ff.

69 Theodor W. Adorno, *Philosophy of New Music*, trans. and ed. Robert Hullot-Kentor (Minneapolis, MN: University of Minnesota Press, 2006), 37; see Adorno, *Prisms*. Adorno regarded Marie Pappenheim's monodrama as "dilettante in its language and structure" (163).

70 Adorno, *Prisms*, 163; see Adorno, *Aesthetic Theory*, 37 ff.

71 Adorno, *Aesthetic Theory*, 181. That "expressionism must accept aspects that reach beyond the τόδε τι" also comprise its realization. For Adorno, *Erwartung*'s gestures of shock assume a formulaic quality once they are repeated. See Adorno, *Philosophy of New Music*, 41 ff.; Adorno, *Aesthetic Theory*, 207.

72 Adorno, *Minima Moralia*, 247; cf. Daniel K. L. Chua, "Drifting: The Dialectics of Adorno's *Philosophy of New Music*," in *Apparitions: New Perspectives on Adorno and Twentieth-Century Music*, ed. Berthold Hoeckner (New York: Routledge, 2006). New music, Chua writes, is "a note of *hope* in the face of despair" (1). Adorno claims that in aiming at oblivion, new music takes upon itself "all the darkness and guilt of the world" (Adorno, *Philosophy of New Music*, 102). As the "true message in the bottle" (102), new music thus remonstrates against the historical catastrophes of the modern age. Chua identifies the expression of suffering in Schoenberg's late twelve-tone structures with music's ghostly return to the presubjective origin symbolized by the sirens' song (see Theodor W. Adorno and Max Horkheimer, *Dialectic of Enlightenment*, trans. John Cumming [London: Verso, 1979]). Accordingly, the negative dialectical cast of the surviving message of a shipwrecked humanity invokes humanity's hope for freedom through bearing witness to the subject's abject alienation. See Daniel K. L. Chua, *Absolute Music and the Construction of Meaning* (Cambridge: Cambridge University Press, 1999). Chua's critical history of the conflicting discourses of absolute music, each of which constructs its meaning in order to exercise its power, trades on the notion that absolute music's "signs signify nothing" (4). For him, the "sign and the referent cancel each other out in such a frictionless economy that no concept of object is left over" (4). Absolute music's alleged purity is therefore "not a fact … but a secret whose power resides in the inaccessibility of its sign" (5). On this view, absolute music's secular eschatological function proves to be a necessary social fiction as the vehicle for the "promise of Utopian happiness" (164). Accordingly, the turn toward "the aesthetic as consolation for a deferred Utopia" (247) provided a refuge for the unfinished project of modernity's emancipatory aspirations.

73 Adorno, *Negative Dialectics*, 6.

74 Ricoeur, *Time and Narrative*, vol. 3, 193.

75 Ibid., 202.

76 Ibid., 198.

77 Ibid., 203.

78 Ricoeur, *Time and Narrative*, vol. 3, 208 ff. See Fredric Jameson, "The End of Temporality," *Critical Inquiry* 29, no. 4 (2003). Jameson remarks on the impossibility of escaping the present's situatedness between the past and the future. For him, the "proliferation of theories of the body … and the valorization of the body and its experience as the only authentic form of materialism" (713) is symptomatic of the reduction of the present to the body's material features. Jameson consequently criticizes "the materialist emphasis on the body today as being fully as ideological as [the] timid spiritualism" (173) that he identifies with the present's metaphysical backing.

79 Paul Ricoeur, *Figuring the Sacred: Religion, Narrative, and Imagination*, trans. David Pellauer, ed. Mark I. Wallace (Minneapolis, MN: Fortress Press, 1995), 208. Hence, "[t]here is nothing new in absolute knowledge; it merely concludes the reconciliation already at work in the successive phases of the philosophical process between certitude and truth" (208). Correlatively, the "end is not something that could be awaited or expected—it is the eternal present of thought that sustains the history of thought.… Absolute knowledge is this divine thought that Aristotle called *noēsis noēseths*, the thought of thought" (208).

80 Ricoeur, *Figuring the Sacred*, 211; see my "The Wager of Imagination and the Logic of Hope," in *Paul Ricoeur in the Age of Hermeneutical Reason: Poetics,*

Praxis, and Critique, ed. Roger W. H. Savage (Lanham, MD: Lexington Books, 2015).

81 Ricoeur, *Figuring the Sacred*, 206.

82 Paul Tillich, *The Courage to Be* (New Haven, CT: Yale University Press, 1952); see Paul Ricoeur, *Reflections on the Just*, trans. David Pellauer (Chicago, IL: University of Chicago Press, 2007), 196.

83 Hoeckner, *Programming the Absolute*, 198.

84 Adorno, *Philosophy of New Music*, 30.

85 Hoeckner, *Programming the Absolute*, 209. Hence Hoeckner's diagnosis: "Liberation occurs at the price of loneliness" (209).

86 See note 68 above.

87 Paul Ricoeur, *Freedom and Nature: The Voluntary and the Involuntary*, trans. Erazim V. Kohák (Evanston, IL: Northwestern University Press, 1966), 453; see Hoeckner, *Programming the Absolute*. Hoeckner comments that both *Erwartung* and Georg Banada's *Ariadne auf Naxos* "make thematic the struggle of the abandoned individual at the 'dawn' and the 'dusk' of the self in bourgeois culture. In both *Ariadne* and *Erwartung*, exterior action is stripped away in favor of an inner drama" (199–200).

88 Bloch, *Principle of Hope*, vol. 3, 1090. As Leonora frees Florestan from his chains, the rising "star of fulfilled hope in the Now and Here" (1103) illumines the dawn of humanity's "morning red, militant-religious" day. *Fidelio* thus stands at the frontier of every storming of the Bastille.

89 Bloch, *Principle of Hope*, vol. 3, 1091. Bloch stresses that: "Atonal music sought to preserve tension in the form of sheer catastrophes" (1094).

90 Bloch, *Principle of Hope*, vol. 3, 1368.

91 Thomas Mann, *The Story of a Novel: The Genesis of Doctor Faustus*, trans. Richard and Clara Winston (New York: Alfred A. Knopf, 1961), 45. Mann attributes this insight to Adorno, who demonstrated to him that

> [h]owever necessary it may be objectively to subject music to rigorous rational analysis, … the effect is just the converse of rationality. Over the head of the artist, as it were, the art is cast back into a dark, mythological realm.
>
> (45)

These consequences adumbrate the dialectic of Enlightenment's reversal (see Adorno and Horkheimer, *Dialectic of Enlightenment*). According to Mann, the

> music, insofar as the novel treats of it (for, to be sure, the novel also practices it) … the music was only foreground and representation, only a paradigm for something more general, only a means to express the situation of art in general, of culture, even of man and the intellect itself in our so critical era. A novel of music? Yes. But it was also conceived as a novel of the culture and the era.
>
> (41–42)

See Hoeckner, *Programming the Absolute*. Hoeckner remarks that *Doctor Faustus*'s allegorical tale of the life of Adrian Leverkühn "stands—*pars pro toto*—for the history of German culture" (226) and its destruction during the Second World War.

92 Thomas Mann, *Doctor Faustus: The Life of the German Composer Adrian Leverkühn as Told by a Friend*, trans. John E. Woods (New York: Vintage International, [1947] 1999), 257.

93 Mann, *Story of a Novel*, 64.

94 Mann, *Doctor Faustus*. By forcing expression from the rigors of twelve-tone constructions, Leverkühn's *Lamentations* performs a "formal negation" (514) of Beethoven's *Ninth Symphony*. See Adorno, *Aesthetic Theory*. Adorno places the dialectic of mimesis and construction at the center of his aesthetic theory. His anthropogenic concept of mimesis thus sets the subject's identification with nature against the anamnesis in art of a condition of freedom that does not yet—and might never—exist. Blind to the causes of unspeakable suffering, art's secularization of this magic identification with nature is accordingly the refuge of art's truth, which recalls this condition of freedom "at the price of [any] real reconciliation" (52). The semblance of freedom in art lies at the root of art's melancholic condition. Consequently, the "survival of mimesis, the nonconceptual affinity of the subjectively produced with its unposited other, defines art as a form of knowledge and to that extent as 'rational.'" Hence, "Schoenberg's dodecaphonic technique ... put into question the principle of development from which it first arose" (Theodor W. Adorno, *The Culture Industry: Selected Essays on Mass Culture*, ed. Jay M. Bernstein [London: Routledge, 1991], 63). See also Hoeckner, *Programming the Absolute*. Hoeckner asks how,

> if Mann imagined Leverkühn's final work as the final stage of the dialectic process between the moment of expression and construction, ... could it be of "utmost calculation" and "*at the same time* purely expressive"? How could the total order of the material come with a "perfectly conscious control over all the 'characters' of expressiveness which have ever been precipitated in this history of music"?
>
> (240; original emphasis)

95 See Bloch, *Principle of Hope*, vol. 3. Bloch champions the "*linear polyphony* of the old counterpoint, ... especially therefore the *fugue*" (1095; original emphases) as the anticipatory model of utopian fulfillment here and now.

96 Mann, *Doctor Faustus*, 512; cf. Bloch, *Principle of Hope*, vol. 3, 1095 ff.

97 Mann, *Doctor Faustus*, 515; see Hoeckner, *Programming the Absolute*. Hoeckner points out that although Mann followed Adorno's advice "not to let Leverkühn leave the cantata unfinished, the concluding 'symphonic Adagio' nevertheless ends like a fragment" (251).

98 Mann, *Doctor Faustus*, 514. Following the death of his nephew, Echo, Leverkühn announces that the "good and the noble ... what people call human, even though it is good and noble.... it ought not to be. It will be taken back. I shall take it back" (501). In the novel, Zeitblom explains that "in Leverkühn's final and loftiest creation, ... echo, that favorite device of the baroque, is frequently employed to unutterably mournful effect" (501). Cf. Olivier Mongin, *Paul Ricoeur* (Paris: Seuil, 1994). Ricoeur in his remarks on Schoenberg's *Moses and Aaron,* which is also a twelve-tone work, highlights the enigmatic character of a mode of expression in which the incommunicability of the unrepresentable figures thematically (241–242). I want to thank George Taylor for bringing this text to my attention.

99 Mann, *Doctor Faustus*, 514.

100 See Mann, *Doctor Faustus*.

> [T]o the very end, his dark tone poem permits no consolation, reconciliation, transfiguration. But what if the artistic paradox, which says that expression, the expression of lament, is born out of the construct as a whole, corresponds

> to the religious paradox, which says that out of the profoundest irredeemable despair, if only as the softest of questions, hope may germinate? This would be hope beyond hopelessness, the transcendence of despair—not its betrayal, but the miracle that goes beyond faith.
>
> (515)

See Hoeckner, *Programming the Absolute.* "Adorno's 'remembrance of otherness' is the ultimate goal of Leverkühn's last work—though no longer as reconciliation…, but as lamentation" (238).

101 Adorno, *Negative Dialectics*, 408.

102 Ricoeur, *Time and Narrative*, vol. 1, 72.

5 Figure of solicitude

Upon receiving word of the death of his friend Charles MacCabe, the blind harper Turlough Carolan composed the lament *The Parting of Companions* (*Scarúint na gCompánach*). Carolan learned about MacCabe's passing from MacCabe, who had dressed himself as a peasant before meeting Carolan on the road near the Fenagh churchyard:

> Thus disguised, MacCabe passing by him after a careless manner addressed Carolan in the usual country salutation of "*Go mbeannuighe Dia dhíbh*" ["God bless ye"]. Carolan returned the salutation of "*Gonn é dhuit*" ["The same to you"], and began to ask about the news of the country. The subtle MacCabe replied that the only news which was in this quarter that the people regretted was the death and burial of a jovial pleasant strolling harper who amused the old and the young, the rich and the poor of the neighbourhood. Carolan enquired of him where he was buried or what was his name. MacCabe replied that he was buried in Fenagh churchyard and that he went by the name of "*Cathaoir an tsiania*" ["Charles the pleasant"]. "Alas, Alas!" cried Carolan, "is he really dead?" "Yes, and buried too", said MacCabe, confirming it with an oath. "Could you show me his grave?" said Carolan. "That I can, as I am the very person who left the last stone over him." "Here is a piece of money", said Carolan, "and bring me to his grave, that I may pay him my last tribute of esteem and respect." MacCabe brought him accordingly over his own pretended grave. Then Carolan desired him to withdraw for a few moments and to attend him when he would call. When Carolan found himself alone he indulged his grief and gave way to his tears. Carolan thereupon composed his "Lament for MacCabe."[1]

The salutary tribulation to which Carolan's plaint gives voice reverberates with the grief that is wrought by the loss of a cherished friend. MacCabe's practical joke notwithstanding, *The Parting of Companions* stands as witness to the sorrow that such parting brings. The weight of this sorrow registers the sadness of our mortal condition, which is ever and again subject to the work of time. Carolan's lament is thus as much a complaint against time's wearing effects as it is a eulogy in praise of his (supposedly) departed fellow harper and compatriot.

By turning to lament, I propose to augment my reflections on music, time, and its other by relating the work of mourning to the grief that is ceaselessly reborn from the passing of our days, years, and lives. In their own ways, death and eternity are limits against which we measure the brief span of our stay on earth. Death and eternity are accordingly also the limits under which we think about time. Over the course of the preceding chapters, I have indicated some of the ways that music sounds out these limits through taking the measure of the difference between time and eternity. In my reflections on Bach, I identified the presentiment of abiding grace with his music's crowning note of joy; in joy, we embrace the good that we desire with respect to the part that we have in being. Conversely, in my reflections on Schoenberg, I highlighted how freeing dissonance from its tonal anchorages unleashes an aporia that by analogy puts radically into question the oneness of time on the historical plane. *Erwartung*'s expressive *tour de force* voices this aporia by reflecting the temporal dispersion of fragmented moments that register the modern alienated individual's psychic distress. In this chapter, I want to search out further the connection between the sublimation of time's deficiency (as in my discussion of the feeling of height in Bach's music) and the imperfect mediation that I set against the aporia of the oneness of time (as in my critique of the "bad infinity" of an instant frozen between the past and the future) within the context of a meditation on lament's cathartic power.

The strategy that I adopt in this chapter has a twofold aim. First, I take up a theme that I left in abeyance when, in my discussion of Bach, I related Pietism's preoccupation with sin and death to the anxiety stemming from the knowledge of one's own death. (This anxiety, I indicated in Chapter 3, is the negative underside of the joy that we have in being.) In considering how the performance of lament aids in working through feelings of grief, I intend to relate the sublimation of the sorrow that accompanies this anxiety, which stems from the awareness of death's inevitability, to the work of mourning. This sorrow, which is the emblem of time's deficiency vis-à-vis our mortally finite condition, opens the way to a broader consideration of a philosophical anthropology of the capable human being (which I will explain in due course). Second, I propose to extend my reflections on the way that working through feelings of grief renews our affinities with the world and with the possibilities that the present holds in readiness by considering the political resonances of some Irish laments. Complaints against political injustices give voice to the hope of deliverance. In turn, this hope fuels the demand for freedom and ignites struggles for recognition. Complaints against injustice reverberate with the longing for an exodus from captivity. As such, they renew the force of the present in answer to the violence of the victor's claim to the right to rule. It will be the task of the final chapter to explore how music's mimetic refiguration of time's ultimate unrepresentability bears on the force that the present has with respect to the ethical and political task of preserving the tension between the spaces of our experiences and the horizons of our expectations. By relating lament's performative role to the politics of lament, my reflections on the work of mourning thus constitute a crucial staging ground for my subsequent investigations into music's power to renew the force of the present in answer both to the violence of history and to the fragility of our mortal finitude.

Mortuary rituals and the "merry wake"

Public performances of poetic lamentations are often part of the traditional fabric of life. In Ireland, public displays of "popular ancestral religious feelings"[2] played a vital role in healing the social and psychic wound that the death of one of its members inflicted on the community. The "merry wake" consequently fulfilled a ritual function that was at once both secular and sacred in restoring the community to health.[3] Like other traditional ritual keening and mortuary rituals that reconcile the community to its loss, this public cultural expression of sorrow and grief institutionalized the work of mourning.

Historically, the practice of waking the dead through ritual keening and revelries in Ireland met with official ecclesiastical opposition. The Catholic Church attempted to stamp out the perceived excesses of the "merry wake" through a number of edicts. From the seventeenth to the twentieth century, synodal meetings of bishops issued regulations that were intended to eradicate keening and "other funeral practices considered to be unchristian, unseemly and a danger to public and private morality."[4] The edict issued by the *Synod of Armagh* (1614) condemned the "performance of 'obscene' songs and 'suggestive' games"[5] as inappropriate even to legitimate contexts of merrymaking. Statute 20 of the *Synod of Tuam* (1660) censured women who engaged in excessive lamentation and keening; the *Synod of Armagh* (1660) went even further by recommending that those who indulged in the merriments of music and dance be excommunicated. The Church was aided in its program of rooting out practices it condemned by the post–Great Famine (1845–1850) emergence of a conservative Catholic middle class. By consciously modeling itself on a "Protestant elite with strong Victorian values, for whom traditional funerary rituals involving lively wakes and lamentation for the dead were an embarrassment,"[6] this conservative class reinforced the Church's opposition to ritual practices that were rooted in, and that contributed to, preserving a traditional way of life.

The waking practices to which the Catholic Church remained opposed throughout the post-Reformation period not only played a vital part in traditional funerary rites, but they were also a symbol of an autochthonous Irish culture's resistance to the Church's ecclesiastical authority. According to Gearóid Ó Crualaoich, the riotous license sanctioned by the "merry wake" became a focus for the carnivalesque element of social life in opposition to modernizing forms of civil and clerical control.[7] The perceived offense to then contemporary Christian sensibilities sprang from the intermingling of Christian and Celtic cosmologies. Deaths that were untimely—as in the case of young children—were attributed to fairy abductions and were part of the magical realm of an ancestral Celtic order. In contrast, deaths that were timely—as in the case of an elderly member of the community who had lived a full life—were regarded as natural and understood to be the result of the will of God.

As a vernacular form of resistance to the Church's official ecclesiastical authority, the mortuary ritual in pre-Famine rural Ireland enacted a familiar assembly custom that was especially prominent in Irish traditional culture at *Samhain* [Halloween], when "otherworld powers, ancestral and fairy, present the greatest danger to human life."[8] As a proximate *Samhain* assembly, the "merry wake" operated at the liminal interface of life and death, where the deceased passed into an afterlife that was "ambiguously ancestral/fairy and Christian."[9] Through staging social and religious feelings belonging to a symbolic order beyond the Church's control, the mortuary rituals of the "merry wake" preserved a sense of continuity with an autochthonous culture, the ancestral cosmological values of which were increasingly undermined by new forms of clerical and social control. Accordingly, the *borekeen* (the trickster-master of ceremonies) and his willing assistants personified the carnivalesque affirmation of the community's cultural and spiritual vitality in the "face of mortally threatening contact with the supernatural realm."[10]

In accompanying the deceased to the grave, the highly charged performance of the *bean chaointe*—the keening woman—also played a crucial part in this traditional mortuary ritual. John Millington Synge's account of the funerary procession that he observed in the Aran Islands at the turn of the twentieth-century underscores the keening women's dramatic role in giving voice to the community's grief:

> After Mass this morning an old woman was buried.... To-day, before the hour for the funeral, poteen was served to a number of men who stood about upon the road, and a portion was brought to me in my room. Then the coffin was carried out sewn loosely in sailcloth, and held near the ground by three cross-poles lashed upon the top. As we moved down to the low eastern portion of the island, nearly all the men, and all the oldest women, wearing petticoats on their heads, came out and joined in the procession.
>
> While the grave was being opened the women sat down among the flat tombstones, bordered with a pale fringe of early bracken, and began the wild keen, or crying for the dead. Each old woman, as she took her turn in the leading recitative, seemed possessed for the moment with a profound ecstasy of grief, swaying to and fro, and bending her forehead to

> the stone before her, while she called out to the dead with a perpetually recurring chant of sobs.
>
> All round the graveyard other wrinkled women, looking out from under the deep red petticoats that cloaked them, rocked themselves with the same rhythm, and intoned the inarticulate chant that is sustained by all as an accompaniment.
>
> The morning had been beautifully fine, but as they lowered the coffin into the grave, thunder rolled overhead, and hailstones hissed among the bracken.
>
> In Inishmaan one is forced to believe in a sympathy between man and nature, and at this moment, when the thunder sounded a death-peal of extraordinary grandeur above the voices of the women, I could see the faces near me stiff and drawn with emotion.
>
> When the coffin was in the grave, and the thunder had rolled away across the hills of Clare, the keen broke out again more passionately than before.[11]

The *bean chaointe*'s ancestral authority as the agent of the deceased's transition to the otherworld underscores her cultural link with the supernatural roles of female figures in a mythic cosmology. According to Ó Crualaoich, the *bean chaointe* is the structural adjunct of the *banshee*, "who in Irish tradition is the supernatural messenger … who is heard crying and keening in the vicinity of the wake-house-to be."[12] The mythic power of the *banshee* harkens back to the great queens that in mediaeval Gaelic literature were associated with royal lineages. Angela Bourke additionally stresses that by claiming that the "Virgin was the originator of the lament tradition[,] … women in Ireland asserted their right to take charge of the death ritual in the face of strong opposition from the Catholic clergy."[13] Public expressions of grief by keening women were therefore at the same time Christian emblems of Mary's iconic embrace of her dead son as in, for example, Michelangelo's Vatican *Pietà*.

This comingling of autochthonous rituals and Christian rites not only afforded a site of resistance to the Church's ecclesiastical and political authority; the "ambiguous Christian/ancestral otherworld of Irish tradition,"[14] to which poetic expression of grief aided the deceased's passage in order that she join her family ancestors, also integrated Irish mythic and Christian cosmologies within the ritual performance of the waking tradition. The *pietas* that Ricoeur argues unites the living and the dead in a "generational model of conscience"[15] is manifest in the performance of waking rites. The familial piety and affection put on display in mourning the dead transcend the ordinary bounds of time and space. Disputes and physical confrontations over whether a deceased woman should be buried with her husband's family or her own attest to the strength of this familial bond.[16] Accordingly, the ritual performance of mortuary rites fulfilled a religious function that was at once both Celtic and Christian.

This *pietas* counterpoints the Christian hope of eternity. Eliade points out that whereas the "Christian liturgy unfolds in a historical time sanctified by the incarnation of the Son of God,"[17] the ritual return in pre-Christian religions to a time *in illo tempore* renews the sacred order of the cosmos. At the same time, Eliade emphasizes how participants in a religious festival consecrate its sacred character through the ritual repetition of cosmogenic acts (as evidenced by different religious rites). The space in which a ritual takes place is set apart from the profane order; the sacred time of the ritual regenerates the temporal rhythms of everyday life. The liminal boundaries of the festival time of the "merry wake" and keening thus not only mark out the difference between ordinary time and the ritual time of its renewal. The carnivalesque activities licensed by the "merry wake" also play a part in healing the community through displaying the vital force that Eliade identifies with the regenerative power of traditional religious rites. Together with ritualized "wailing and groaning,"[18] which punctuates displays of public mourning, the festival time of the "merry wake" authorizes a period of Celtic-Christian revelry.[19] The dual role of the "merry wake" in promoting the carnivalesque catharsis of feelings of grief and in reestablishing the social order complements the *bean chaointe*'s role as the symbol and agent of the process that enables the deceased to be incorporated among her familial ancestors. The performance of mortuary rites fulfills the living's obligation to the dead in renewing the time of the living. This mortuary tradition's efficacy is at once both socio-political and sacred. In aiding the deceased in her passage to her Christian-ancestor world, the observation of traditional mortuary rites is thus at the same time the ritual staging ground for the work of mourning.

The work of mourning

Poetic laments for the dead are a testament to the depths of human suffering and grief. The therapeutic value of funerary practices is at once both cosmic and existential in this respect. As I outlined above, the ritual performance of traditional funerary rites heals the wound inflicted on the living through fulfilling the familial obligations that unite the living and the dead across the generations. At the same time, these ritual practices provide a time and a space for renewing the time of the living in response to the loss of one's own. Expressions of sorrow and suffering to which laments give voice acquire their therapeutic value in the context of this work of mourning. As such, these expressions of grief are also the affective index of our fragility and vulnerability as mortal human beings.

This affective index registers the absence of aseity evidenced by the contrast between time and eternity. As such, expressions of suffering and grief counterpoint the abilities that we have as capable human beings. The sorrow that accompanies the knowledge of our mortal finitude is therefore not the final word. The fragility expressed in the vocabulary of the inability or the lesser ability to speak, to tell our own stories, to act in accordance with our convictions and beliefs, and to hold ourselves accountable to others is the negative underside of our powers and capacities. Our ability to exercise these powers and capacities is never absolute. Like the fragility of our identities, which is rooted in the struggle to think for ourselves in opposition to another's hegemonic rule, our vulnerabilities are indices of regions of experience where our being affected attests to different centers of alterity.[20] Setting the fragility of identity against the presumption of autonomy that Ricoeur emphasizes is the "prerogative of the subject of rights"[21] not only underscores how the violence of a history of colonization, for example, institutionalizes the belief on the part of subjugated populations of their lesser abilities and powers but also constitutes a critical transit point in political and social struggles when tied to a philosophical anthropology that expressly thematizes the powers that we have as capable human beings. This philosophical anthropology foregrounds the threats stemming from (1) identity's susceptibility to its ideological deformation due to the split in the concept of identity between *idem* identity and *ipseity*, or *ipse* identity; (2) others' ways of living that contest or undermine our ways of leading our lives; and (3) the heritage of founding acts of violence that are legitimated after the fact by narrative justifications of the victor's right to rule.[22] These threats attest to the vagaries of history and time. Consequently, a philosophical anthropology that sets our power to exercise our capacities and abilities against our vulnerability and fragility underscores the temporality of our impermanent condition in contrast to eternity's anteriority, exteriority, and/or height.

While laments sometimes consist of wailing and moaning, elaborate poetic and musical compositions that are remembered and sung for generations also have a place in mourning rituals. In Ireland, the improvised art of lament constitutes a poetic and song genre that is part of the Irish language tradition. The English term "keen" has its origin in the Irish word for this genre: *caoineadh*.[23] According to Breandán Ó Madagáin, the keen as practiced in Ireland "could be a genuine instrument for the heightened expression of personal emotion."[24] Each of the three parts that make up a round of keening thus has a crucial function. In the salutation, the "mourner … commences by some deep murmuring, repeating over and over the name of the deceased such as '*A Thomáis, A Thomáis, mo chumha is mo dhíth thú*' ['O Thomas, O Thomas, my sorrow and my loss!']."[25] (Calling out to the dead is one of the most powerful expressions of bereavement. Hence the lament is "almost always a direct address to the dead person."[26] According to folk belief in some parts of Ireland, the deceased retained the power to hear until the priest threw "three shovels of earth on the coffin at burial."[27]) The second part, the verse or dirge, was sung in Irish in "praise of the deceased."[28] The third part, the choral "cry" [*gol* or *gáir*], which concluded each verse, was led by the keening women, who were joined by others present in singing vocables such as *och ochone* or *ululoo*.[29] By using vocables rather than words, the "community gave poignant expression to their emotion in purely musical terms, using their voices as a musical instrument, just as in the instrumental lament played on the pipes or the harp."[30]

The sorrow that in the round of lamenting described by Ó Madagáin culminates in a communal performance of a vocables song is emblematic of the "sadness of the finite"[31] that inheres in our mortal condition. All suffering attests to the pathos of our finite existence through laying bare the limits of our capacities and powers, as I indicated in my earlier remarks on our fragility and vulnerability. Ricoeur reminds us that we must be cognizant of the more deeply concealed forms of suffering for which no expression is adequate: "the incapacity to tell a story, the refusal to recount, the insistence of the untellable—phenomena that go far beyond mishaps and adventures, which can always be made meaningful through the strategy of emplotment."[32] Vocables seemingly give voice to a grief for which words are lacking. As such, these expressions of grief echo the unsayable and the unspeakable through sounding out the depths of sadness and despair.

In her reading of some famous Irish laments—including the "Lament for Art Ó Laoghaire" (*Caoineadh Airt Uí Laoghaire*) by his wife Eibhlín Dubh Ní Chonaill (Eileen O'Connell)—Angela Bourke shows how laments enact the stages of grieving that Elisabeth Kübler-Ross describes in *On Death and Dying*.[33] Poetic expressions of denial, anger, bargaining, sadness, and acceptance mirror the process of grieving. Denial, for instance, is evinced by the "common motif whereby the lamenter calls on the dead man to get up and come home."[34] Anger, bargaining, sadness, and finally acceptance mark out lament's performative trajectory. The lamenter facilitates her listeners' expressions of grief through moving them to tears. As such, the lamenter plays the part of a "kind of grief therapist"[35] within a poetic genre, the power of which sets it among the most enduring utterances within the Irish tradition.

Expressions of anger seem at first to be contrary to lament's solicitous expressions of sorrowing sadness. The "obverse of the praise and celebration of order which are found in all laments,"[36] anger gives vent to an overwhelming sense of powerlessness. Similar to feelings of disregard that ignite struggles for recognition, feelings of helplessness inflame the sense of indignation that fuels protests against the brevity of life, the hard hand of fate, or social and political injustices. In her "Lament for Art Ó Laoghaire," Eibhlín Dubh Ní Chonaill—aunt of Daniel O'Connell "The Liberator"—denounces Abraham Morris for murdering her husband. Under the Penal Law of Ireland in force at the time, Protestants were authorized to demand that Irish Catholics sell them their horses for £5.5s. Ó Laoghaire, whose chestnut mare had beaten Morris's horse at the Macroom races, refused to accept the £5 that Morris had offered him. The quarrel between Morris and Ó Laoghaire escalated and, according to commonly held belief, Ó Laoghaire was outlawed. On May 4, 1773, Ó Laoghaire was shot dead on his horse.

Eibhlín's accusation against Morris is biting and swift:

May you lose your sight
May your legs be broken
For killing my true love
And not a man in Ireland
Dares to shoot at you.[37]

The demand for justice is rife with bitter political experiences:

For Jesus Christ knows well
I won't cover my head
I'll wear no shift
I'll go barefoot
No wares in my home
The brown mare uncontrolled
But I'll expend to lawyers
I'll sail over the sea
For an audience with the King
And if he won't hear me out
Then I'll return to Ireland
To turn on that black-blooded scoundrel.[38]

Ó Laoghaire—the "last person shot as an outlaw in Ireland"[39]—became a symbol of English injustice. Although a coroner's jury found Morris guilty of the crime in June 1773, Morris was never punished. Receiving no justice from English courts, Eibhlín Dubh Ní Chonaill had her kinsmen kill Ó Laoghaire's murderer.

Praise and blame: These two warring epithets lie at the heart of lament's performative task. On one side, an accusation directed against the cause of one's grief. On the other side, a litany of respect and honor that constitutes the eulogy of love. The balance between them must be carefully struck.[40] For where blame raises its voice against the wrong, praise celebrates the memory of the one whose absence is the reason for another's grief.

The evil of wrongdoing and the evil of suffering, however, belong to two heterogeneous categories. As for wrongdoing, human conduct is blameworthy where the action undertaken violates the prevailing moral codes. Here the harm inflicted by one on another bears the stamp of the evil of violence. As for suffering, the evil that is endured is one that is not merited by the conduct or actions on whom this evil is inflicted. Lament, Ricoeur therefore emphasizes, is in a sense also the opposite of blame: "whereas blame makes culprits of us, laments reveals us as victims"[41] of another's conduct. Here, the evil committed by one is the cause of the evil suffered by another. Ricoeur emphasizes that the "translation of the German *Klage* by *lamentation* in French and by lament in English does not capture the provocative kinship in German between lament [*Klage*] and accusation [*Anklage*].'[42] According to him, the cry of lamentation is thus most acute at the point of their intersection.

The enigmatic reversal of blame into praise attests to lament's performative power. Through interrogating the cause of one's suffering, the lament fixes the affective dimensions of feelings of abandonment and grief in its poetic expression. In turn, the meditative mood to which the accusation gives way leads from the abyss of grief to the heights of praise. The tone of Ní Chonaill's supplication for her dead husband to return follows fast on her words of honor and acclaim:

> My beloved friend
> How well a hat suited you
> With a band of gold on it.
> A sword with a hilt of silver
> And the haughty prancing
> Of your horse with the white blaze
> English people used to bow
> Down to the ground in front of you
> Not from love, but from abject fear
> And yet you died at their hands....
> My dear love and sweetheart
> Rise up, stand up
> And we'll go home.[43]

The image that in Ní Chonaill's lament personifies Ireland's maternal tenderness for her fallen son casts its desolate shadow over Ireland's political landscape and thus turns the accusation against British injustice toward the work of mourning:

> My stalwart friend
> Who belonged to the noblest of the Uí Laoghaire
> To Art from Gougane Barra
> And Tadhg the irascible one with the breastplates
> Of the bright, resolute hero
> Who was my husband
> Till destiny struck
> Out on the hill, with no priest
> .... I found you, stretched out, dead
> Like a common animal
> Under a gorse bush
> With neither pope nor bishop
> Nor clerk nor priest
> Who would pray over you
> Except for a poor destitute woman
> Who had thrown her cloak over you.[44]

Grief, Ricoeur reminds us, is "that sadness that has not completed the work of mourning."[45] Like plaints against the evil of suffering, laments have a place in this work through giving voice to the sorrow of our loss. Mourning tends toward melancholia when, succumbing "to the blows of its own devaluation,"[46] the ego abases itself. Ricoeur goes so far as to ask whether "reproaches addressed to the self serve to mask reproaches aimed at the loved object."[47] Melancholy is therefore more than just a psychic disturbance. For "once we begin to consent to sadness, to fatigue, to discouragement,"[48] melancholia gives way to despair. Poetic laments, mortuary rituals, and funerary rites resist the complacency toward the sadness of *acedia*, "which the spiritual leaders of the Middle Ages saw [as] the worst of temptations."[49] By staging feelings of grief, these rites, rituals and laments thus combat this complacency toward sadness, which threatens to overwhelm us whenever the death of a loved one brings us face to face with the vulnerability and fragility of our mortal state.

Feelings of grief consequently echo the sadness of meditative memory that renders the "'mood' of finitude … conscious of itself."[50] This sadness, for which there is no cause, constitutes the ultimate vis-à-vis of the work of mourning. By likening this melancholic sadness to Kierkegaardian sickness-unto-death—the "relative of despair"—or Gabriel Marcel's notion of "no hope (*inespoir*),"[51] Ricoeur gives this "mood" of finitude its fullest existential weight. This "mood" registers the deficiency marked by the difference between the brevity of our mortal lives and an order of time that outstrips them. In response to this deficiency, the "evocation of a sublime sadness"[52] in music, poetry, literature, and visual works of art evinces a piece that Ricoeur tells us is missing from the panoply of Freud's metapsychology. The reversal from a "complaisance toward sadness to sadness sublimated"[53] crowns the work of mourning. Joy is accordingly "the reward for giving up the lost object"[54] in being reconciled to one's loss. This reconciliation lies at the heart of lament's cathartic power. The sublimation of feelings of grief is thus the demonstration and proof of lament's performative role in ritual practices that renew the order of the living in answer to the brevity of our fleeting stay.

The Parting of Companions

Turlough Carolan's lament for his friend Charles MacCabe is a testament to the way that feelings of grief resonate with the sadness of our mortally finite condition. The opening verse is a plaint against the bleakness of a destiny that awaits us all:

> *Nach truagh sin mise agus mé tuirseach i ndiaidh mo shiúil*
> *Ar uaigh mo charad agus mé ag folcadh na ndeor go buan*
> *Ní bhfuair mé agam is é do lagaigh ar radharc mo shúl*
> *Ach crualeac dhaingean agus leaba den chré bhí cúng.*[55]

How pitiable am I, weary after my travels,
At my friend's grave, ceaselessly shedding tears
I found nothing there – it has weakened my eyesight [further]-
But a hard unyielding flagstone and a narrow bed of clay.[56]

The misery symbolized by the grave's unyielding hold stands as witness to an inconsolable grief. The imagery of this mournful landscape reverberates with the "musicality of the poetic word."[57] The accompanying musical strain amplifies the sorrowing sadness of this wretched scene. Through sound and word, Carolan's lament gives voice to the grief that death's partings bring (Example 5.1).

Example 5.1 Turlough Carolan, *The Parting of Companions.*[58]

However much aesthetics has been regarded as superfluous to studies of Gaelic cultural practices, the "expression of feeling in the performance of song"[59] is critical to *The Parting of Companion*'s cathartic force. Lillis Ó Laoire, who is himself a nationally recognized *sean-nós* singer from Donegal, underscores the importance of evoking feelings in a successful song performance. Feeling's poetic transposition renders raw emotions, the visceral immediacy of which are often overwhelming, communicable. The "structuring of sound, rhyme, rhythm, intonation, assonance, and so on, furnishes the stabilizing factors that haul back and bring to a stand the fleeting word that points beyond itself."[60] Gadamer consequently explains that the "constantly shifting balance between sound and sense"[61] crystallizes in images whose affective tonalities reverberate by reason of them. Like a tune or a piece of music, the mood evoked by a poem is coextensive with the affective resonances of its internal order. Hence, like music, a poem or a lyric text evokes feelings and moods that it alone expresses.

According to some accounts, Carolan generally composed the music for his airs before writing the accompanying text. One harper, for example, told Edward Bunting that "Carolan always made the tune first and the poem last."[62] Lesa Ní Mhunghaile suggests that since the "air determined the metre, … the words were deliberately chosen to match the notes."[63] O Sullivan similarly remarks that "Carolan's verse is often extremely irregular in structure, the irregularities corresponding to the irregularities of phrasing which are a leading characteristic of his music."[64]

The verse that embodies the "inevitable pun on … [Charles MacCabe's] Christian name"[65] suffuses Carolan's lament with its elegiac tone of kingly praise:

> *Ni cathaoir mar chathaoir an Cathaoir fó 'gcaoiniom,*
> *Ach Cathaoir ba taitneamhaighe ná cathaoir na ríoghtha.*
> *Níor shuidh Cathaoir i gcathaoir ó chaitheamh na díle*
> *Ba taitneamhaighe nó mo Chathaoir, 'sé Cathaoir a' tsiamsa é'*
>
> The cathaoir whom I mourn is not a cathaoir like others
> But a more radiant Cathaoir than the cathaoir of the kings
> No Cathaoir sat on a cathaoir since the time of the deluge
> More pleasant than my Cathaoir, Cathaoir the entertaining one.[66]

The comparison with the splendor of the "chair of kings" stands as a testament to the strong evaluation that draws Carolan's praise of MacCabe into the orbit of the discourse of *agape*. The reciprocal relation that in this discourse takes the form of an entreaty addressed to the beloved to love in return here assumes the tenor of a salutary tribulation. Friendship, Ricoeur reminds us, is *the* unique good. The confidence that takes root in the feeling of a mutual solicitude thus forges the bonds that unite friends, family, and lovers. This confidence is a feature of the dialectic between "binding and unbinding that is common to both the speculative and the empirical features of love."[67] While unbinding "speaks of the suffering of absence and distance, … binding speaks of the strength of spirit that is embodied in the capacity to be alone."[68] Hence, only the "confidence in the permanence of a reciprocal solicitude … makes such unbinding a salutary tribulation."[69] The sublimation of sadness that in *The Parting of Companions* is a token of the bereaved's reconciliation with the other's loss is at the same time a ritual of the soul. Like an incantation that dissolves all distance, the air's sublime grace invokes the joy of a "happy" memory on the horizon of the work of mourning through summoning this joy in the midst of the air's sorrowing plaint, as for example, in Matt Molloy's performances of *The Parting of Friends*.[70]

Charles MacCabe, whose prank prompted Carolan to compose *The Parting of Companions*, was not in attendance at Carolan's funeral. However, according to O'Sullivan, it was fitting that MacCabe, "and no other, should write the elegy of a man whose life's experience had marched so closely with his own."[71] Carolan was not only MacCabe's "fellow-harper, and boon companion, whose genius he was probably better able to appreciate than most of Carolan's contemporaries,"[72] but MacCabe's plaint for Carolan is also a lasting testament to the reciprocity of the salutary tribulation born from the feeling of a mutual solicitude. Lamented by Carolan, MacCabe laments in return:

> *Nach í so an chuairt easbhach a laguidh mé réis mo shiubhail,*
> *Ar uaigh mo charad is do mharbh sin radharc mo shúl.*
> *Ní bhfuair mé 'falcadh na mdeór go h-úr,*
> *Ach cruaidh-leac dhaingean is leabuidh don chré ró-chumhang....*
>
> *Míle is seacht gcéad, ochtdéag is fiche go dian*
> *Ó theacht Mhic Dé dár saoradh ó bhruid na bpian*
> *Nó gur feartadh i gcré an égise b'áirde, clú,*
> *Cearabhlán céillidhe, cémeamhail, ceólmhar ciúin.*
>
> What a futile visit, one that has weakened me after my travels
> To my friend's grave and that has extinguished my sight even more fully
> Although I was shedding fresh tears at every moment
> I found nothing there but an unyieldingly hard flagstone and a narrow bed of clay....
>
> A thousand, seven hundred eighteen and twenty precisely
> Since the son of God came to save us from the bondage of our pain
> Since the poet of high renown was interred in the ground
> Judicious Carolan, esteemed, musical, grave.[73]

Politics and the poetics of lament

Likening the bondage of grief to the pain of captivity from which, according to Christian faith, the Son of God delivers the faithful underlines the latent political force of the hope of deliverance expressed in some Irish laments. Carolan's *The Parting of Companions* secured a place in the history of the Irish struggle against English rule when Edward Bunting performed it for Theobald Wolfe Tone at the Belfast Harp Festival on July 11, 1792. Wolfe Tone—the father of Irish republicanism—was instrumental in founding the Society of the United Irishmen. The Society's objective in forming a political union between Catholics and Protestants soon turned from seeking parliamentary reform to establishing an Irish republic through armed insurrection. Following the trial and suicide of his compatriot the Reverend William Jackson, Wolfe Tone was permitted to emigrate to America in May 1795. In 1798, Wolfe Tone joined the invading French forces in support of the Irish rebellion. Following his capture, he was tried and sentenced to hang on November 12, 1798. Before the sentence could be carried out, Wolfe Tone cut his throat, dying by his own hand on November 19, 1798 at the age of 35 in Provost's Prison, Dublin.

Complaints against political subjugation and social injustice set the demand for freedom against the history of Ireland's colonialist oppression. Political humiliations that sometimes were couched in expressions of personal grief fueled feelings of indignation that ignited the spirit of revolutionary opposition to British rule.[74] The enduring interest in Ní Chonaill's keen for her murdered husband attests to the place that the memory of British injustice has in the popular imagination. Similarly, vernacular histories, stories, and songs memorialize injustices through commemorating collective traumas of local and national events. Local stories and vernacular histories that contest official accounts of the 1798 Rebellion, for example, not only keep alive the memory of the insurrection's heroic struggle, but these so-called counterhegemonic stories and histories also legitimate the nationalist struggle against British authority.[75]

Traumas and grievances stored in the vernacular archives of a collective memory are accordingly living traces of the social injustices and political humiliations endured. Guy Beiner rightly reminds us that history is not so much written by the victors but is instead forgotten by them. Histories that sanction the official celebrations of commemorated events trade on the selective forgetting of the suffering of these histories' victims. Beiner consequently emphasizes that "Irish folk history and the vernacular commemorative practices it cultivated"[76] had an ethical as well as a social and political value. Through reliving events to which stories, songs, and commemorative monuments bore witness, local communities could identify and express their empathy with their ancestors' traumatic experiences. In opposition to "official" histories, folk history and vernacular commemorative practices thus constituted sites of resistance that mark out the landscape for remembering places and events that official commemorative practices largely ignored.

The spontaneous sites of mourning that Beiner remarks coexist with official sites of memory are emblems of these vernacular histories' commemorative function. Traditions of folk commemoration authorized "local communities to remember and honor *their* dead"[77] in ways that were often personal. Laments for rebels who died in the 1798 Rebellion, for example, recall the anguish wreaked on families and friends. For Beiner, the commemorative-psychotherapeutic function of the social memory expressed through the folklore of lament aids the work of mourning. At the same time, the memorialization of embittered sentiments that were officially censured (sentiments that Beiner regards as representing a subaltern *mentalité*) can also repeat the psychic and symbolic wounds stored in the archives of a community or nation's collective memory. Nuala Ní Dhomhnaill remarks that imperialist regimes ultimately triumph only when they prevent the material, physical, and psychic traumas endured by subjugated populations "from being expressed as it need be—in anger and in forgiveness—so that it remains impacted into the psyche of the colonized."[78] For her, facing the "deep level of collective amnesia about who we are and where we come from"[79] is vital to overcoming an existence that she regards as highly fragmented at best and at the worst is pathogenic. In her view, possessing the land "emotionally and imaginatively without any particular sense of, or need for, titular ownership"[80] through *dinnseanchas* (place-lore) provides a kind of therapeutic response to the wound the Famine inflicted on the Irish psyche. For Ní Dhomhnaill, *dinnseanchas* accordingly inscribe the memory of this collective trauma on the landscape through the

> pointing out of the graves and its delineation of certain spots as "hungry grass," where famine victims died, and where if you walk on them, you will be overcome by a sudden and overpowering hunger and weakness that might even be lethal.[81]

By focusing on those who fought the Rebellion and lost, local folklore not only preserved the memory of traumatic incidents and events written out of the official record, but, as acts of remembrance that countermanded the acts of forgetting that were strategic to the victors' narrative self-legitimation, provincial folk history also resisted official accounts of the Year of the French, in which vernacular histories had no place. Just as crucially, *seanchas*—"shorter realist accounts"[82] through which social and historical traditions were passed down orally—situated the present in relation to the past recounted in local stories. Perhaps more crucially still, commemorative sites and the stories recounted played and play a vital part in the continuing "creation of the future out of the past."[83] The present, Henry Glassie points out, is "a gift from the past"[84] that holds as yet unfulfilled hopes and expectations in readiness. For, as I pointed out earlier, the present acquires its force as the time in which we exercise our capacity to affect history's course only insofar as we preserve the tension between the horizons of the past, which have already been surpassed, and the horizons of as yet unfulfilled demands and aspirations. Glassie explains that "At home and in the public house, in their place of small hills and holy warfare, the people of Fermanagh construct their past, telling the simultaneous story of their community and their nation,"[85] and in so doing claim their future as their own. Histories recounted in stories and songs thus preserve the memory of rogues, heroes, and ordinary folk caught up in the struggle for Ireland's liberation; like a song's accompanying *údar*, which vests the singer and the song with their authority, the truth of these histories takes hold in the storyteller's and his or her listeners' identification with them.[86]

The Suffering Ducks March is a testament to the power of the living history of stories and songs. Sam Murray, one of Ireland's foremost traditional flute makers, learned the tune from his father's lilting on an Appalachian dulcimer that his sister had sent over from the States. Sam's father was a member of the Soprani Accordion Band in the York Road area in Belfast, which played this tune as an anthem in memory of the forced expulsion of the Catholic community from the surrounding loyalist area during the 1920s Troubles. The members of this community, which was an offshoot of the Sailortown, Belfast docks workers, were nicknamed the "suffering ducks."[87] The tune, the title of which Cairon Carson describes as ruefully ironic, memorializes experiences that are part of a living history.[88] As such, the tune and the story that accompanies it preserve the memory of a community whose place in the history of sectarian strife might otherwise have been erased (Example 5.2).

Example 5.2 The Suffering Ducks March.

The evaluate texture of stories that are told and recounted invariably draws the memory of past events into the orbit of the present. Narrative manipulations exploit the selective operation of remembering and forgetting to justify the victors' claims to the right to rule. As the "guardian of identity,"[89] ideology thus mobilizes memory in response to the demand for identity. This selective operation molds the contours of the community's understanding of its history, its prospects, and its own self-worth. Consequently, stories that celebrate the glory of founding events or victorious battles too often promulgate denigrating images that, when internalized, redouble the harm suffered by those subject to another's domination. The grip on memory of commemorative acts, Ricoeur accordingly emphasizes, is the "apanage of all those enamored of glory."[90] Conversely, in contrast to a selective forgetting on the part of those who write the history of the victors, too much memory tends toward the compulsion to repeat, which "leads us to substitute acting out for the true reconciliation by which the present would be reconciled with the past."[91]

The abuses that Ricoeur identifies with memory's narrative manipulation trade on the selective strategy that forges the compact between forgetting and remembering. This strategy turns the narrative imposed in this way into the privileged instrument of a way of judging the actors in the story that is told. Heroes and victims, agents and patients, protagonists and antagonists are all a function of the plot. To tell the story is thus to judge the history of which one is a part. The exemplary value of traumas recounted in vernacular histories turns the memory of the past toward future expectations (as Ní Dhomhnaill's remarks attest). The poetic discourse of praise and blame that in lament joins the work of remembering to the work of mourning thus acquires its specifically political accent by vesting the protest against injustice with the hope of deliverance's moral force.

The prophetic character of *aisling* poetry clearly accentuates the prospective dimension of protests against social and political injustices. In the Irish tradition of *aisling* poetry, the moral riposte to the harm inflicted takes the form of a prophetic vision of Ireland's political liberation. Ireland's personification as a young or old woman in eighteenth-century *aisling* vision poetry is replete with allegorical references to Ireland's struggle for independence. Beiner explains that the "Jacobite *aisling* (vision/dream) poetic genre … centered on a legendary female character—the *spéirbhean* (sky-woman), who originally assumed a royal persona."[92] In lamenting the decline of Gaelic and Catholic influence and power, *aisling* poetry held out the promise of Ireland's overthrow of British rule. The *Shan Van Vocht* (Poor Old Woman), for example, draws on this tradition in proclaiming imminent victory in the song "The French are on the Sea!" [93] In this Jacobite tradition, the woman personifying Ireland often brought a "message of hope, that the Stuarts will return and restore the old order."[94] Even though the Stuart cause was no longer a real possibility by the end of the Seven Years War (1763), "Jacobite formulae, with their archaic references to Bonny Prince Charlie and the Catholic powers, continued to be adapted to articulate popular grievances about rent, tithe, the new police, or the activities of Bible Societies by songsters dependent on popular support, right up to the Famine."[95] Consequently, the millennialist message of liberation conveyed by the genre of *aisling* poetry was one with which every social class could continue to identify.

Aisling poetry's prophetic character thus serves as a kind of counterweight to the complaint against political injustices and the sufferings wrought by them. The allegorical figure of the woman who brings a message of hope is the same figure that in Ní Chonaill's lament for her murdered husband Art Ó Laoghaire covers his body with her cloak. This intimate gesture of tenderness symbolizes the feeling of solicitude evoked through the poetic imagery of Ireland's maternal care. Prophecies of deliverance that turn plaints against political captivity toward the visionary horizon of a land of freedom thus renewed the hope of Ireland's cultural and political restoration.

In his St. Patrick's Day speech in 1934, Éamonn De Valera—Ireland's first *taoiseach* (prime minister)—set out his own vision of the new nation as the

> home of a people who valued material wealth only as a basis of right living, … [and of] a land whose countryside would be bright and cosy homesteads, … whose firesides would be the forums of the wisdom of serene old age.[96]

This vision for a return to Ireland's autochthonous Celtic values has an eschatological ring that echoes in plaints against the brevity of life and the power of time to destroy. Together, prophetic poetry and laments for the dead mark out regions of expressions of grief where plaints against sorrow and suffering intermingle with the joy of an insouciant freedom on the horizon of the work of mourning.

The longing for an exodus from exile in one's own land consequently reverberates in the lament for a captive nation. The Irish poem "The Deserted Land" evokes the desolate landscape of English oppression and injustice out of which the Jacobite dream of deliverance sprang. Accordingly, "The Deserted Land" stands as pledge of the strength of the spirit to endure:

> There is no laughter at a child's deeds, music
> ceases, Irish is in chains....
> No praise poem is recited, no bedtime story told,
> no desire to see a book, no giving ear to the family
> pedigrees....
> The captivity that was in Egypt has overtaken them....
> .... how shall the oppression be lifted from
> the bright fair-haired race of Conn, since we have no
> Moses in Ireland?[97]

A deeper memory of exodus suffuses this plaint. In return, praise in the form of the memory of banished pleasures and joys fuels the longing for freedom from captivity.

This reference to a deeper memory of exodus marks a further turning point in my reflection on music, time, and its other. When, in the next chapter, I set music's simulations of a time beyond time against the backdrop of an anthropology of the mythic return to a time *in illo tempore*, I will distinguish between music's fictive explorations and the sacred character of myth. Apocalyptic and eschatological resonances of representations of the "end of time" preserve vestiges of this mythic return to the *illud tempus*. Yet, far from renewing the order of the cosmos, these aesthetic simulacra refigure time's ultimate unrepresentability in accordance with the ways that they allow time to be transcended from within. The deeper memory of exodus that in "The Deserted Land" renews the hope of deliverance thus marks the transition in my inquiries from the poetics and politics of lament to a consideration of some of the ways that music revalues time in response to the contrast between time and eternity that emerges with the break separating history from myth.

Notes

1 Donal O'Sullivan, *Carolan: The Life Times and Music of an Irish Harper* (Cork: Ossian, 2001 [1958]), 49. O'Sullivan remarks that there are many versions of the practical joke that MacCabe played on Carolan. The one

> told by Bartholomew O'Dowd is perhaps as good as any, and it has the advantage of giving a location to the incident—the churchyard of Fenagh.

> Fenagh is a village north of Mohill, County Leitrim, through which Carolan would pass when proceeding from his home in Mohill to visit his patrons the Maguires, who lived at Tempo, in County Fermanagh.

MacCabe afterward disclosed his identity and "rallied the good-natured bard on his giving such a sincere proof of his affection for one who had so often made him the butt of his wit" (Joseph Cooper Walker, *Historical Memoirs of the Bards: Interspersed with Anecdotes of, and Occasional Observations on the Music of Ireland. Also, An Historical and Descriptive Account of the Musical Instruments of the Ancient Irish. And an Appendix Containing Several Biographical and Other Papers, with Select Irish Melodies.* London: Printed for T. Payne and Son at the Mews-Gate, and G. G. J. and J. Robinson, Pater-Noster-Row [1786], quoted in O'Sullivan, 49).

2 Gearóid Ó Crualaoich, "The 'Merry Wake,'" in *Irish Popular Culture, 1650–1850*, eds. James S. Donnelly, Jr. and Kerby A. Miller (Dublin: Irish Academic Press, 1998), 175. Crualaoich explains that the custom of assembling together was especially prominent in Irish cultural tradition. See Patricia Lysaght, "'Caoineadh os Cionn Coirp': The Lament for the Dead in Ireland," *Folklore* 108 (1997). According to Lysaght, this public ritual, which "appears to have been part of the fabric of Irish life a thousand years ago," (66) continued into the twentieth century. See also Angela Bourke, "The Irish Traditional Lament and the Grieving Process," *Women's Study International Forum* 11, no. 4 (1988). Bourke points out that the "Irish practice of lamenting the dead is mentioned in several early texts and was remarked on by travelers from the time of Elizabeth I" (288).

3 Ó Crualaoich, "Merry Wake," 191. In Ireland, the "American wake" was similarly held for a family member or friend prior to her or his departure for similar reasons. Faced with the likelihood of never seeing a loved one again, bidding farewell to a friend or family member who was leaving the land of her birth was akin to mourning her passing.

4 Lysaght, "Caoineadh," 66; see Ó Crualaoich, "Merry Wake." The *Maynooth Synod* (1927), for example, continued to outlaw "the practice of engaging in immodest behaviour in the presence of the corpse" (174).

5 Ó Crualaoich, "Merry Wake," 174.

6 Lysaght, "Caoineadh," 68.

7 Ó Crualaoich, "Merry Wake," 173. Ó Crualaoich provides a detailed description of traditional funerary rites. The waking tradition provided a public venue for expressing feelings and values rooted in an ancient, mythic past in opposition to official ecclesiastical authority.

8 Ó Crualaoich, "Merry Wake," 191.

9 Ibid.

10 Ibid., 193.

11 John Millington Synge, *The Aran Islands* (Boston, MA: John W. Luce and Co., 1911), 63–64.

12 Ó Crualaoich, "Merry Wake," 192.

13 Bourke, "Irish Traditional Lament," 290:

> Hundreds of song texts have been collected which represent the Virgin Mary lamenting her son at the crucifixion, and in these religious songs, sung for the most part by thoughtful and devout women, the Virgin is described in exactly the same terms as the lamenter in the secular tradition: she loosens her hair and runs barefoot through the wilderness; she screams at the soldiers who

> are crucifying Christ and she leaps high into the air "like a madwoman."... Singers explain that it is right to lament the dead in the traditional way, because the Virgin Mary did so. Sometimes they say that it was she who invented the custom.

14 Lysaght, "Caoineadh," 69. Lysaght comments that the performance of keening was also an opportunity for the bereaved family to enhance its standing in the community by

> ensuring that the deceased was generously lamented. The wake for the dead, anchored for so long in the home setting in Ireland, offered the family one of its best opportunities, perhaps, to exhibit its wealth and possessions to relatives and community.
>
> (67)

See Ó Crualaoich, "Merry Wake." Ó Crualaoich similarly emphasizes the ambiguity of a worldview that is at once both Christian and Celtic (176).

15 Paul Ricoeur, *Oneself as Another*, trans. Kathleen Blamey (Chicago, IL: University of Chicago Press, 1992), 354. Ricoeur points out that this "generational model of conscience contains ... [an] impenetrable enigma: the figure of the ancestor, beyond relatives whether close or distant, begins a movement of infinite regress in which the Other progressively loses ... the initial, presumed familiarity."

16 See Ó Crualaoich, "Merry Wake":

> explanations of death as fairy abduction are tied to the incorporation process in Irish tradition. Clearly, a young wife dying in childbirth and subsequently regarded as the victim of fairy abduction has not been fully or finally incorporated into her husband's family life. Thus her own family were especially concerned to bury her body in their own graveyard near their own ancestors as the nearest they could come to reincorporation and recovery from the otherworld of the sí.
>
> (195–196)

Ó Crualaoich subsequently provides an account of a burial dispute (198).

17 Mircea Eliade, *The Sacred and the Profane: The Nature of Religion*, trans. Willard R. Trask (Orlando, FL: Harcourt, 1959), 72; italics removed.

18 Bourke, "Irish Traditional Lament," 287; see Lysaght, "Caoineadh." Lysaght describes how lament marks salient moments in the waking ritual (74 ff.)

19 See Eliade, *The Sacred and the Profane*. Eliade explains that the

> abolition of profane past time was accomplished by rituals that signified a sort of "end of the world." The extinction of fire, the return of the souls of the dead, social confusion of the type exemplified by the Saturnalia, erotic license, orgies, and so on, symbolized the retrogression of the cosmos into chaos. On the last day of the year the universe was dissolved in the primordial waters.
>
> (78–79)

This periodic regression annihilated everything that "time had soiled and worn.... By symbolically participating in the annihilation and re-creation of the world, man too was created anew; he was reborn, for he began a new life" (79).

20 See the Introduction, note 16.

21 Paul Ricoeur, *Reflections on the Just*, trans. David Pellauer (Chicago, IL: University of Chicago Press, 2007), 73.

22 See Paul Ricoeur, *Memory, History, Forgetting*, trans. Kathleen Blamey and David Pellauer (Chicago, IL: University of Chicago Press, 2004), 80 ff.; see also my "Fragile Identities, Capable Selves," *Études Ricoeuriennes/Ricoeur Studies* 4, no. 2 (2013).

23 Lysaght, "Caoineadh." Improvised poetic lament by women was "highly developed and persisted well into the twentieth century" (65); see Bourke, "Irish Traditional Lament," 287.

24 Breandán Ó Madagáin, "Irish Vocal Music of Lament and Syllabic Verse," in *The Celtic Consciousness*, ed. Robert O'Driscoll (Toronto: Dolmen Press, 1981), 312.

25 Frank Keane, cited from manuscript 12 Q 13 (10) in the Royal Irish Academy by Ó Madagáin, "Irish Vocal Music of Lament," 312; italics added.

26 Bourke, "Irish Traditional Lament," 288.

27 Lysaght, "Caoineadh," 69; see Ó Crualaoich, "Merry Wake," 183.

28 Ó Madagáin, "Irish Vocal Music of Lament," 312. Ó Madagáin cites Bunting, who in *Ancient Music of Ireland* (Dublin 1840) "gives an example of the music to which the verse was sung, adding the note 'sung by a single voice in praise of the deceased'" (312).

29 Ó Madagáin, "Irish Vocal Music of Lament," 313. Ó Madagáin points out that the English reference to "the Irish cry" refers to this choral cry.

30 Breandán Ó Madagáin, *Caointe agus Seancheolta Eile/Keening and Other Old Irish Musics* (Indreabhán: Cló Iar-Chnnacht, 2005), 84; quoted in Sean Williams and Lillis Ó Laoire, *Bright Star of the West: Joe Heaney, Irish Song Man* (New York: Oxford University Press, 2011), 96.

31 Paul Ricoeur, *Fallible Man*, trans. Charles A. Kelbley (New York: Fordham University Press, 1986), 140.

32 Ricoeur, *Oneself as Another*, 320; see 190 ff.

33 Bourke, "Irish Traditional Lament," 288. According to Bourke, "[o]ne of the most important insights offered by the work of Elisabeth Kübler-Ross is that the normal process of grieving always includes some element of anger and rage" (289); see Elisabeth Kübler-Ross, *On Death and Dying* (London: Tavistock, 1973).

34 Bourke, "Irish Traditional Lament," 288. According to Bourke, some of Ireland's most notable laments "were composed for men who died suddenly or violently" (289).

35 Bourke, "Irish Traditional Lament," 288.

36 Ibid.," 290.

37 Translation by Lillis Ó Laoire. For an Irish version of the lament, see Angela Bourke et al., eds., *The Field Day Anthology of Irish Writings*; Volume IV: *Irish Women's Writings and Traditions* (Cork: Cork University Press, 2002). See also Louis M. Cullen, "The Contemporary and Later Politics of *Caoineadh Airt Uí Laoire*," *Eighteenth Century Ireland/Iris an dá cultúr* 8 (1993). Cullen maintains that the lament is not a true *caoineadh*. According to him, some of the verses were written at a later date than the funeral, so that "we are not dealing with a *caoineadh* written at the time of death, … [but] with a composition, which, whatever its poetic form, is not strictly speaking a *caoineadh*" (10). See Seán Ó Coileáin, "The Irish Lament: An Oral Genre," *Studia Hibernica* 24 (1988); Seán Ó Coileáin, "Oral or Literary? Some Strands of the Argument," *Studia Hibernica* 17/18 (1977/1978); Lysaght, "Caoineadh." Lysaght explains that in "its most developed form in Ireland, the traditional lament or keen was a long extempore poem of eulogy and sorrow composed in *Rosc* metre" (71). According to her, the "The Keen for Art O'Leary" was "first written down from the recitation of a County Cork keening woman in 1800, and again from the same woman prior to her death in 1873." It appeared in print toward the end of the nineteenth century. The lament was probably "composed in sections at a number of locations over a period of time" (72). Eibhlín likely

> performed the initial, most personal and intimate, part of the lament at the spot where her husband had died.... A second phase of lamentation occurred during the wake in the O'Leary home prior to the removal of the body for burial.... Art was given temporary burial, probably because of a decree of the Penal Laws prohibiting burial within monastic precincts...; and between this burial and his final interment in the O'Leary cemetery, Eibhlín added to her lament, giving public vent to rage and grief, naming and cursing those responsible for her husband's death and calling for revenge.... As he was finally laid to rest with his ancestors she performed her final public session of lamentation. It also appears that she added some stanzas to the lament in later years.

38 Translation by Ó Laoire.

39 Maurice O'Connell in a letter to Crofton Crocker (1834), quoted in Cullen, "Contemporary and Later Politics of *Caoineadh Airt Ui Laoire*," 24. Cullen points out that this claim by Daniel O'Connell was penned by his son Maurice O'Connell. Cullen argues that the lament's significance is due in large part to the political resonances of the events leading to Ó Laoghaire's murder. Morris supported the Shannon political interest that "had driven the anti-catholic campaign both in the county in the 1760s and in the 1762 and 1764 sessions in parliament" (15–16). The "publicly-expressed support for Morris in 1771–73 was exclusively drawn from the Shannon interests in the county and city" (16). The tragedy of Art Ó Laoghaire, like that of Morty O'Sullivan, is thus "part of the larger story of the place Catholics held in the most important political county in Ireland" (18–19). The final stage of the Ó Laoghaire saga was a result of the events between 1834 and 1844, with the addition of the inscription referring to the English government's role in Ó Laoghaire's demise on his tombstone:

> Lo! Arthur O'Leary,
> Generous, handsome, brave
> Slain in his youth
> Lies in this humble grave
> Died May the 4th 1773 aged 26 years
>
> (Cullen, 36)

40 See Bourke, "Irish Traditional Lament," 209.

41 Paul Ricoeur, *Figuring the Sacred: Religion, Narrative, and Imagination,* trans. David Pellauer, ed. Mark I. Wallace (Minneapolis, MN: Fortress Press, 1995), 250:

> The whole enigma of evil may be said to lie in the fact that, at least in the tradition of the West, we put under the same terms such phenomena as sin, suffering, and death.... One principal cause of suffering is the violence human beings do to one another. In fact, to do evil is always, either directly or indirectly, to make someone else suffer.

42 Paul Ricoeur, "Lamentation as Prayer," in *André LaCocque and Paul Ricoeur, Thinking Biblically: Exegetical and Hermeneutical Studies*, trans. David Pellauer (Chicago, IL: University of Chicago Press, 1998), 214. Ricoeur highlights the paradoxical formulation of this cry in his exegesis of Psalm 22:

> On the one side, the lament [in Psalm 22] comes close to being an accusation [of having been "abandoned by God,"]; on the other side, it remains caught up within an invocation and held within the space of prayer, inasmuch as it is addressed to God.
>
> (218)

See Ricoeur, *Memory, History, Forgetting*, 73.

43 Translation by Ó Laoire.

44 Ibid.
45 Ricoeur, *Memory, History, Forgetting*, 77.
46 Ibid., 73.
47 Ricoeur, *Memory, History, Forgetting*, 73: "Freud proposes the hypothesis that the accusation, by weakening the object-cathexis, facilitates the retreat into the ego, while transforming the discord with the other into self-laceration."
48 Ricoeur, *Reflections on the Just*, 196. Ricoeur stresses that this threat inheres in each of us. Conversely, the

> contrary to this consent to sadness is that moral sentiment, or better, that spiritual attitude Paul Tillich called the "courage to be." The courage to be unites in one braid the work of memory, the work of mourning, and self-esteem.
>
> (196)

See note 82, Chapter 4.
49 Ricoeur, *Memory, History, Forgetting*, 76.
50 Ibid.
51 Ibid.
52 Ricoeur, *Memory, History, Forgetting*, 77. Ricoeur hears in Beethoven's last quartets and sonatas the "powerful evocation of a sublime sadness."
53 Ricoeur, *Memory, History, Forgetting*, 77.
54 Ibid.
55 Tomás Ó Máille, *Aṁráin Ċearḃalláin: Poems of Carolan Together with Other N. Connacht and S. Ulster Lyrics* (London: Irish Texts Society, 1916), 162.
56 Translation by Ó Laoire.
57 Hans-Georg Gadamer, *The Relevance of the Beautiful and Other Essays*. ed. Robert Bernasconi, trans. Nicholas Walker (Cambridge: Cambridge University Press, 1986), 134.
58 Edward Bunting, *A General Collection of Ancient Irish Music, Containing a Variety of Admired Airs never before Published and also the Compositions of Conolan and Carolan, Collected from the Harpers & c. in the different Provinces of Ireland, and adapted for the Piano-Forte*, vol. 1 (Dublin: W. Power & Co., 1796), 13 [No. 25]).
59 Lillis Ó Laoire, "Towards an Aesthetic of Song Performance," *Folk Life: Journal of Ethnological Studies* 54, no. 1 (2016), 53. Ó Laoire, who twice won the *Corn Uí Riada* at *Oireachtas na Gaeilge* (in 1991 and 1994), emphasizes the significance of aesthetics in this regard. He recounts that in a television interview he conducted with Connall Ó Dónaill—a "*seanchaí*, singer and cultural and language activist born in Rann na Feirste, Rannafast, C. Donegal in 1911" (50)—Ó Dónaill told of a song he learned from his great aunt. His great aunt, Ó Dónaill remarked, "refused to accept anything but a performance that was emotionally engaged" (51). Ó Laoire adds that under his own apprenticeship to Ó Dónaill, he himself experienced the "challenge of touching the emotional core of any given piece and of transmitting that to others." See Williams and Ó Laoire, *Bright Star of the West*. Williams and Ó Laoire explain that the

> word *fonn* in Irish—to describe a sung melody—is borrowed from the Latin *fundus*, the base or fundament. *Fonnadóir* and *fonnadóireacht*—terms for singer and singing, respectively, in the Irish of the Carna area—were, until recently, used only locally.... The *fonnadóir* proceeds from the basic theme of a song and works with it in myriad subtle ways, not just through melismatic ornamentation, but through variation, transference and reversal of melodic motifs, and a type of gentle play between poetics and sound.
>
> (109)

60 Gadamer, *The Relevance of the Beautiful*, 134.
61 Ibid.
62 Quoted in Lesa Ní Mhunghaile, "Carolan's Verse," *Amhráin Chearbhalláin/The Poems of Carolan: Reassessments*, ed. Liam P. Ó Murchú (Dublin: Irish Texts Society, 2007), 53. According to Ní Mhunghaile, the practice of "composing the air first" was not unusual.
63 Ní Mhunghaile, "Carolan's Verse," 53. For Ní Mhunghaile, "Carolan's attempts at matching lyrics with airs appears to have met with varying success" (53). Carolan, who was born during the Restoration period of modern Ireland's cultural creation, sometimes has been described posthumously as an Irish bard and hence as belonging to the last of a vanishing race. Ní Mhunghaile, however, points out that "with the decline of the bardic schools the role of the bard merged with that of musicians, so that by the eighteenth century harpers fulfilled both functions" (52). See Nicholas Carolan, "Carolan's Music" in *Amhráin Chearbhalláin/The Poems of Carolan: Reassessments*, ed. Liam P. Ó Murchú (Dublin: Irish Texts Society, 2007), 3.
64 O'Sullivan, *Carolan*, 87.
65 Ibid., 49.
66 Translation by Ó Laoire: "Cathaoir an tsiamsa also means Cathaoir the fun-loving – lover of entertainment especially musical entertainment" (Lillis Ó Laoire, pers. comm., August 4, 2016).
67 Paul Ricoeur, *The Course of Recognition*, trans. David Pellauer (Cambridge, MA: Harvard University Press, 2005), 190.
68 Ricoeur, *The Course of Recognition*, 190.
69 Ibid.
70 "Matt Molloy-Air: The Parting of Friends/Paddy Ryan's Dream," YouTube video, 5:27, posted by "Jeremy V," January 1, 2012, www.youtube.com/watch?v=3_uSji7Azz4&list=RD3_uSji7Azz4#t=97, accessed September 15, 2016; "Irish Flute: Matt Molloy Plays a Slow Air and Reels," YouTube video, 5:03, posted by "Douglas Hadden," May 13, 2015, www.youtube.com/watch?v=SiBUPFBi8Tc, accessed September 15, 2016. See Gadamer, *The Relevance of the Beautiful*, 162; Ricoeur *Memory, History, Forgetting*, 77; see also Chapter 3.
71 O'Sullivan, *Carolan*, 67.
72 Ibid.
73 "Marbhnadh Uí Carolan-Mac Cabe's Elegy on Carolan." The Irishpage.com, accessed September 26, 2016, http://irishpage.com/songs/carolan/elgycarl.htm; translation by Ó Laoire. According to O'Sullivan, while the "poem exists in many versions, … that printed in Connellan's *Duanaire (Fonna Seanma)* … is incomparably the best" (O'Sullivan, *Carolan*, 67). The title given by Connellan is "Charles MacCabe's Lamentation for his Dear and Beloved Friend, Terence Carolan, at his Grave in the Church of Kilronan, County Roscommon."
74 For a discussion of the struggle for recognition, see Ricoeur, *The Course of Recognition*; Axel Honneth, *The Struggle for Recognition: The Moral Grammar of Social Conflicts*, trans. Joel Anderson (Cambridge, MA: MIT Press, 1995). The commemoration of events erased by "official" histories fueled this revolutionary struggle by keeping alive the memory of social and political humiliations.
75 See John Hutchinson, *The Dynamics of Cultural Nationalism: The Gaelic Revival and the Creation of the Irish Nation State* (London: Allen and Unwin, 1987).
76 Guy Beiner, *Remembering the Year of the French: Irish Folk History and Social Memory* (Madison, WI: The University Wisconsin Press. 2007), 304.
77 Beiner, *Year of the French*, 203.
78 Nuala Ní Dhomhnaill, "A Ghostly Alhambra," in *Irish Hunger: Personal Reflections on the Legacy of the Famine*, ed. Tom Hayden (Boulder, CO: Roberts Rinehart Publishers, 1997), 69; see Williams and Laoire, *Bright Star of the West*, 73.

79 Ní Dhomhnaill, "Ghostly Alhambra," 69.
80 Ibid., 77.
81 Ibid.
82 Beiner, *Year of the French*, 82. *Seanchas* has a variety of meanings: "history, lore, ancient law, a record or register, a minute description, a pedigree, an ancient tale; [an] act of storytelling, gossiping; inquiring (about one's condition, health, etc.)" (81). Beiner comments that the content of these traditional accounts could also be fanciful. *Seanchas* differ from *scéalaíocht*, which are structurally complex folktales that were often versions of international ones.
83 Henry Glassie, "Tradition," *The Journal of American Folklore* 108, no. 430 (1995): 395.
84 Henry Glassie, *Passing the Time in Ballymenone: Culture and History of an Ulster Community* (Philadelphia, PA: University of Pennsylvania Press, 1982), 143.
85 Henry Glassie, *Irish Folk History: Tales from the North* (Philadelphia, PA: The University of Pennsylvania Press, 1982), 17.
86 See Hugh Shields, *Narrative Singing in Ireland: Lays, Balls, Come-All-Yes and Other Songs* (Dublin: Irish Academic Press, 1993):

> Songs therefore are celebrations more than entertainments and as such may claim attention for their communicative value first and afterwards for their artistic effect. The confirmatory "*údar*", whether told or not told, exists or must be invented. Those who listen to songs know that certain things occurred which are the occasion of the songs, even if they do not know what things they were.... [T]he things told are often fictional.... Yet they cannot be doubted.... [,] particularly when the changing identities of the singing "narrator" merge into [the] "I" [of the song] and the ambiguities of past action and action unrealized are resolved by assigning all action to the region of sentiment.
>
> (83)

See Glassie, *Passing the Time*: "True stories commence in identification" (131). Glassie notes that "[e]very story is made to fit circumstances, but if the truth abiding apart in a timeless realm of absolutes does not appear, unscathed and intact, during performance, the tale is falsehood" (144).
87 Sam Murray, pers. comm., November 26, 2010. Murray recounted that he attended a school in York Road that had been built for this community. See Cairon Carson, "The Suffering Ducks: How Has a Belfast Flute Sound Evolved in Such a Short Time?," *The Journal of Music* (April 1, 2009), accessed September 20, 2016, http://journalofmusic.com/focus/suffering-ducks. Carson remarks that the story behind the tune brings another dimension of meaning to it. Sam Murray, Carson adds, "is himself perhaps a kind of Suffering Duck" who left Belfast for reasons not entirely unrelated to the sectarian strife in the area.
88 Carson, "The Suffering Ducks." According to Carson, the tune is a "jaunty march, almost a joyful swagger." Harry Bradley and Michael Clarkson play the tune on the album *The Pleasure of Hope* "on shrill band flutes pitched in F full of attack and crack, and you could almost take it as an Orange tune."
89 Ricoeur, *Memory, History, Forgetting*, 83.
90 Ricoeur, *Memory, History, Forgetting*, 85–86. The abuses of memory are grafted onto ideology's legitimating and integrative functions. On the deepest level—that of action's symbolic mediation—memory is "incorporated into the formation of identity" (84–85) through the narrative function. Distortions that feed on the symbolic systems that Ricoeur identifies with ideology's constitutive function multiply the abuses that turn collective memories into instruments of power and control. This constitutive function, he points out, can scarcely operate outside of its connection to the justificatory function of a system of power. Furthermore,

this second function cannot operate "even potentially apart from the function of distortion" (83) that is grafted onto it. Ideology's legitimating and dissimulating functions are implicated in this two-fold operation of selective remembering and forgetting. The surplus value—the supplement—that ideology adds to the belief in a ruling authority in response to that authority's claim to legitimacy masks the discrepancy between them. This surplus value's narrative texture thus lends itself to ideological discourses that justify the power exercised by one group over another. Armed with a "history that is itself 'authorized'" (85) by those in power, the commemoration of events deemed to belong to a common history and identity promotes a selective social memory. See Benedict Anderson, *Imagined Communities: Reflections on the Origin and Spread of Nationalism*, rev. ed. (London: Verso, 1991). Anderson singles out the obligatory role that this selective strategy of forgetting plays in reshaping the memory of religious and national conflicts and wars. "Having to 'have already forgotten tragedies' of which one needs to be reminded," Anderson maintains, "turns out to be a characteristic device in the later construction of national genealogies" (201). For Anderson, the systematic historiographical campaign that is deployed in France principally through the state's school system reminds "every young Frenchwoman and Frenchman of a series of antique slaughters which are now inscribed as 'family history.'"

91 Ricoeur, *Memory, History, Forgetting*, 79. The selective strategies of remembering and forgetting are invariably caught up in the paradox that Ricoeur reinterprets in terms of Sigmund Freud's psychoanalytic categories of resistance and the compulsion to repeat.

92 Beiner, *Year of the French*, 92; see May McCann, "Music and Politics in Ireland: The Specificity of the Folk Revival in Belfast," *British Journal of Ethnomusicology* 4 (1995): 55. McCann identifies *Róisín Dubh* as a Gaelic *aisling* song composed in the seventeenth century that identifies Ireland with a woman metaphorically (60). See also Sean Williams, "Melodic Ornamentation in the Connemara Sean-nós Singing of Joe Heaney," *New Hibernia Review* 8, no. 1 (2004): 132; Tomás Ó Canainn, *Traditional Music in Ireland* (Cork: Ossian, 1993), 69 ff.

93 See Beiner, *Year of the French*. According to Beiner, the song is "probably a late-eighteenth century composition about General Hoche's expedition to Bantry Bay in 1796 or General Humbert's expedition to Killala in 1798" (92).

94 Tom Dunne, "Subaltern Voices? Poetry in Irish, Popular Insurgency and the 1798 Rebellion," *Eighteenth-Century Life* 22, no. 3 (1998): 34.

95 Dunne, "Subaltern Voices?," 35.

96 Éamonn De Valera, quoted in Edward O. Henry, "Institutions for the Promotion of Indigenous Music: The Case for Ireland's Comhaltas Ceoltoiri Eireann," *Ethnomusicology* 33, no. 1 (1989): 69; see Hutchinson, *Dynamics of Cultural Nationalism*. Hutchison argues that the image of Ireland "as a holy island—*insula sacra*—blest by nature with peculiar advantages" (55) formed a nucleus for eighteen-century revivalist fervor. The formation of a radical reformist political organization, which Theobald Wolf Tone named the Society of United Irishmen, brought a socio-political dimension to Irish revivalism. Motivated in part by the rejection of English values, the "return to Irish exemplars and a worship of blood, soil and geography" (67) animated nationalist sentiments and aspirations.

97 Aindrias Mac Marcais, "The Deserted Land," in *The Field Day Anthology of Irish Writing*, vol. 1, ed. Seamus Deane et al. (Derry: Field Day Publications, 1991), 280. The poem is also published in the original Irish in this volume of *The Field Day Anthology.*

6 The ends of time

The evocation of a feeling of being "out of time" evinces the reach of music's power. "Eternity" experiences in which a succession of instances gives way to a "standing present" radically transcend the everyday order of time's wearing effects. In such experiences, the sempiternal order of the *nunc stans* supersedes temporal processes of movement and change. Music's explorations into the various ways in which time can be surpassed by its other consequently draw near the enigma of time's ultimate unrepresentability through overtaking the limits of the temporally regulated flow of events. These imaginative forays into regions where time gives way to its other border on myth. Accordingly, time's inscrutable character constitutes the central axis around which various figurations of a "time beyond time" vis-à-vis the return to a time *in illo tempore* gravitate.

By relating music's explorations into the boundaries between time and eternity to the enigma that bears the stamp of the difference between them, I want to highlight the place that representations of the "end of time" have within the panoply of experiences in which temporal succession gives way to a present without a past or a future. The "standing present" of this eternal "now" not only contrasts with the kind of temporal progression typically associated with forward-directed movement; the apparent annihilation of succession and change also sets this eternal present's timeless expanse apart from our everyday ways of reckoning with time. Opposing a static, immovable duration to dynamic processes seems to set radically anti-teleological conceptions of music's temporal character against teleological ones. However, as I indicated in Chapter 2, this opposition obfuscates the reliance of these seemingly mutually exclusive conceptions of musical time on cosmological time. The enigma of time's ultimate inscrutability prevents us from reducing simulacra of a "time beyond time" to representations of an instantaneous "now" point infinitely extended along a rectilinear axis by pushing the confrontation between phenomenological time and cosmological time to its utmost limit. Consequently, music's evocation of a super-temporal reality that exceeds or transcends our ordinary experiences of time refigures time's ultimate unrepresentability in answer to our fundamental inability to constitute or to master time.

My second concern in this chapter complements the first. By focusing on representations of the "end of time," I intend to set my reflections on the ways that music revalues time in response to time's ultimate inscrutability against the backdrop of an anthropology of the return to a mythic time *in illo tempore*. Insofar as representations of the "end of time" in some modern and postmodern works employ stylistic devices that purportedly annihilate any sense of temporal progress, the ethos of "time transcended" in these works sometimes approximates this archaic return to a time of "great beginnings" before the dawn of human history. The difference between these representations of the "end of time" and the mythic return to a time *in illo tempore* rests on the break that separates history's existential valorization from the ritual repetition of cosmogenic acts. In the course of the reflections that follow, I will therefore argue that modern and postmodern variations of time's annulment revalue the significance of the force of the present with respect to history's temporalization. By showing how these "eternity" experiences' eschatological or apocalyptic resonances are mimetic ripostes to the existential condition born from the break that separates history from myth, I will highlight the challenge that these explorations into the being and the nonbeing of time raise with regard to the insuperable difference between our historical situation as mortally finite beings and mythic vestiges of the search for transcendence. The various figures of time's abolishment in modern and postmodern music are irreducible to the return to a sacred time *in illo tempore*. Yet can we not hear an echo of this mythic return in the ways in which different evocations of the feeling of being beyond time seek to overcome both history and time?

Time and its other

The distance separating modern and postmodern representations of the "end of time" from the ritual repetition of cosmogenic acts sets music's mimetic responses to time's inscrutable character against the mythico-religious experience of transcendence. By identifying this mythico-religious experience of transcendence with the ritual renewal of the cosmos, Eliade accentuates the fundamental difference between history and myth. Through reenacting a sacred event that took place in a primordial, mythical time, participants in a religious festival reactualize a sacred time sanctified by the gods. For *homo religious*, sacred time thus "appears under the paradoxical aspect of a circle, … [as] a sort of eternal mythical present that is periodically integrated by means of rites."[1] Refusing to live entirely in the historical present as we conceive it, *homo religious* finds her true being in the eternal repetition of the cosmos's fundamental rhythms.[2]

With the world's demythologization, the eternal recurrence of the cosmos's cyclic creation and destruction seemingly gives way to time's devouring power. The terror of history that Eliade opposes to the renewal of cosmos through the return to the *illud tempus* threatens the sacred order of myth. Where for *homo religious* the ritual repetition of cosmogenic acts undoes the work of time through restoring the cosmos to its original state, the non-religious individual only recognizes the existential dimension of time's wearing effects. Time reveals its unmasterable power through constituting the "deepest existential dimension" [3] of our mortally finite existence. Heidegger, Eliade accordingly remarks, "has gone to the trouble of showing that the historicity of human existence forbids all hope of transcending time and history,"[4] severed as this existence is from a sacred time *ab origine*.

Setting the terror of history against the mythic return to the *illud tempus* sharpens the contrast between time and eternity in a way that underscores the connection between the dimension of passivity that I identified earlier with the alterity of time's other and the vertical dimension of the feeling of height. On the one hand, the experience of being subject to the effects of time bears out the seeming complicity between time's existential constitution vis-à-vis time's distention and dispersion, and a cosmological order that is indifferent to our sufferings, hopes, and fears. The grief that is continually reborn from the disproportion between the brevity of our mortal lives and the heavenly bodies' sempiternal revolution is thus also an emblem of the fundamental element of passivity rooted in the absence of aseity in our historically finite existence. On the other hand, the feeling of height sets this fundamental element of passivity against the feeling of height's verticality through interiorizing the anteriority, exteriority, and superiority of time's other. For Eliade, no world is possible without this vertical dimension, since the "structure of the cosmos keeps [the] memory of the celestial supreme being alive …. as if the gods had created the world in such a way that it could not but reflect their existence."[5] It is an open question as to whether poetic expressions of this vertical relation preserve the trace of this archaic memory. In his reading of Hölderlin's poem, *Der Ister*, Heidegger maintains that the upward glance that spans the dimension between earth and sky gives the measure of our mortal dwelling. Poetry, he therefore insists, is the "primal form of building"[6] inasmuch as it speaks in images of what remains hidden by the blinding, radiant height from which we take the measure of our stay on earth. The hidden god to whom this glance is directed cannot be wrested from her or his concealedness. For the god who remains unknown is only revealed as such by way of the span that bridges between earth and sky. Hence, the "sky is not sheer light."[7] Rather, the "radiance of its height is itself the darkness of its all-sheltering breadth."[8] Thus, the "god who remains unknown is revealed *as* such by the sky."[9]

Like myth, which Eliade emphasizes is the "expression of a *mode of being in the world*,"[10] and poetry, which as Ricoeur stresses intends being, music opens us to the world and the world to us. From this vantage point, the proximity of representations of the "end of time" with the ritual return to a time *in illo tempore* highlights the affinity of "eternity" experiences with mythic time. Eternity's heterogeneity with respect to the existential constitution of time consequently invites a reflection on music's mythopoetic significance. By the same token, the difference between these representations of time's overcoming or annihilation and the ritual repetition of cosmogenic acts sets music's mimetic responses to time's inscrutable character against the mythico-religious experience of transcendence, as I indicated at the beginning of this chapter. It may be that the "experience of efficacy par excellence"[11] is one in which the numinous element of the sacred declares itself in a prelinguistic mode. On Ricoeur's account, hierophanies in which the sacred manifests itself thus "attest to an inscription of the sacred in a level of experience beneath that of language."[12] Hence, this experience can properly be regarded as aesthetic. Music's proximity to myth undoubtedly fuels the temptation to vest music's evocation of a "time beyond time" with a noumenal force. Confusing the sacred's inscription on the level of the aesthetic with music's efficacy, however, risks reduplicating the conceits that in the nineteenth century legitimated the claim to music's spiritual transcendence. Thus, as I have said previously, vesting music's affective power with a metaphysical value only obfuscates music's real significance with respect to time's ultimate inscrutability.

Experiences of transcendence undoubtedly seem at first to justify the idea that music's power to induce ecstatic states shares in the superreal efficacy of hierophanies. Gilbert Rouget, for example, remarks that the material density given by music to time alters our consciousness of states of being to which music alone gives us access. The musical architecture of trance experiences and states of musical possession accordingly structures a superreal order that sanctifies ritual actions or possession states *hic et nunc*.[13] Like trance experiences and ritual possessions, the "logic of religions of transcendence"[14] also depends on music's power to suspend ordinary states of consciousness. Music's superreal efficacy thus plays a crucial role in modifying or transfiguring a manner of "being-in-the-world"[15] through overcoming the ritual fault that separates the sacred order of the divine from the profane order of everyday existence.

The ethnographic evidence of music's superreal efficacy is striking. For John Blacking, music that he regards as intended for being is an "adventure into … the reality of the world of the spirit."[16] Far from providing a means of escape, the experience occasioned by *tshikona* (the Venda national dance) is a marker of Venda identity in transcending its social function. Through creating a "special world of time,"[17] *tshikona* fulfills its ritual function through drawing the members of the community together. The altered sense of time that A. J. Racy identifies as a principal feature of *ṭarab* ecstasy similarly provides ethnographic proof of music's ontological vehemence. The "extraordinary emotional state"[18] that accompanies the sense of temporal transcendence induced by the music intermingles feelings of joy (*farah*) with those of a sublime sadness (*huzun*). *Ṭarab* ecstasy, Racy emphasizes however, differs from mystical states through which union with the Divine is attained. *Ṭarab* is instead related to *salṭanah*, which is an ecstatic state in which the "performer becomes musically self-absorbed (*mundamij*), and experiences well-focused and intense musical sensations."[19]

Rouget's account of trance and possession experiences and Racy's explanation of *ṭarab* ecstasy highlight how music's ritual and mythic value relates to its mimetic significance vis-à-vis time's ultimately inscrutable character. I admit that I may be extending these authors' interpretations in ways that they may not have intended. Yet, taken together, Rouget's remarks regarding music's power to alter states of consciousness and the distinction made by Racy between *ṭarab* ecstasy and mystical-religious states point to a critical difference between music's mythic relation to a ritual fault and music's mimetic power. Music's power to evoke feelings of being out of time are operative in ritual practices that redress the ritual fault that separates the *illud tempus* of the world's creation from the world's profane, temporal order. Conversely, neither this ritual fault nor the mythic return to a time *in illo tempore* has a place in the existential constitution of history or time. However closely modern or postmodern music's evocation of a "time beyond time" emulates the myth of eternal return, the time of history and the time of myth are irreparably divided. For us, the enigma of the ultimate inscrutability of time thus places its limit on the passage beyond, which is the ritual achievement of time's transcendence through the mythic return to a time *in illo tempore*.

The rupture between history and myth has a special place in my reflection on music's significance. Barred from the return to the time of myth by the world's demystification, explorations into the difference between time and its other take the form of fictive variations of "eternity" experiences. These fictive variations, however, have a real value with respect to the significance that the difference between the being and the nonbeing of time has for us. Through staking out the borderlines between time and its other, music's mimetic refiguration of time's ultimate unrepresentability revalues time in accordance with its evocation of an order beyond it. My thesis that music refashions our manner of inhering in the world consequently acquires its larger ambitus in the light of the irreparable separation of myth from history's temporalization.

It may be that "as the terror of history grows worse, ... humanity ... will strive to forget, as meaningless and dangerous, any spontaneous gesture which might entail 'historical' consequences."[20] How otherwise, Eliade asks, could we tolerate the horrors of our historical catastrophes in the absence of any sign of some transhistorical meaning? Conversely, Arendt alerts us to the dangers of suppressing the freedom to intervene in the world's course that is rooted in our specifically human capacities for thinking and acting. For Eliade, however, only the soteriological value of a freedom that "has its source and finds its guaranty and support in God"[21] can justify our experience of history as a meaningful event. Hence, for him, only a philosophy of freedom that reserves a place for God can transcend the horizon of mythic archetypes with impunity.[22] The problematic of freedom's actualization within humanity's historical reality that I will introduce in Chapter 7 thus comes face to face with our ultimate inability to master time. For a hermeneutical reflection that takes time and its other as its theme, the alterity of an order that supersedes the temporal limits of our mortally finite condition thus marks out the horizon of the question of the meaning of time.

Quatuor pour la fin du Temps

> And I saw another mighty angel come down from heaven clothed with a cloud: and a rainbow was upon his head, and the face was as it were the sun, and his feet as pillars of fire.... And he set his right foot upon the sea, and his left foot on the earth.... And the angel which I saw stand upon the sea and upon the earth lifted up his hand to heaven, and sware by him that liveth for ever and ever.... that there should be time no longer; but in the days of the voice of the seventh angel, when he shall begin to sound, the mystery of God should be finished.
>
> Revelation of St John the Divine 10: 1–7

Olivier Messiaen's *Quatuor pour la fin du Temps* is an iconic transcription of the soteriological message inspired by this passage from the Book of Revelation. The work received its première at Stalag VIII A on January 15, 1941, where the Germans had imprisoned Messiaen after his capture during the Second World War. Camp commanders, guards, and prisoners attended the première, which took place in the depths of winter.[23]

For Messiaen, the Catholic mystic and Christian believer, the announcement, that in the days when the trumpet of the seventh angel of the apocalypse is about to sound there will be no more delay until the mystery of God will be revealed, presages the fulfillment of God's divine plan.[24] Père Jean-Rodolphe Kars comments that in his works, Messiaen celebrated the "immense path of the History of Salvation, from Creation, through Redemption up to Consummation ... with unprecedented magnificence."[25] The last movement of Messiaen's *Quatuor pour la fin du Temps* symbolizes the final passage from Time to Eternity. Seven, Messiaen explains, "is the perfect number, the six days of Creation, sanctified by the Divine Sabbath."[26] By extending this Sabbath to eternity, for Messiaen eight becomes the number of "unwavering light, unalterable peace."[27] The eighth movement of *Quatuor pour la fin du Temps, Louange à l'Immortalité de Jésus*, accordingly represents humanity's ascent toward Paradise. This ascent crowns *Quatuor pour la fin du Temps*' quasi-programmatic representation of Eternity's eschatological triumph. Transcribed from the second movement of Messiaen's 1930 organ work, *Diptyque*, *Louange à l'Immortalité de Jésus* evokes the ecstatic consummation of the divine mystery. The solo violin's assured climb to its highest register is the emblem of the individual's ascent toward God. Through this iconic ascension to the most supreme height, Time rejoins Eternity (Example 6.1).[28]

Example 6.1 Olivier Messiaen, *Louange à l'Immortalité de Jésus*, mm. 28–33.
Source: *Quatuor pour la fin du temps.* Music by Olivier Messiaen. Copyright © 1942 Éditions Durand – Paris. Tutti i diritti sono riservati – All rights reserved. Reprinted by permission of Hal Leonard MGB, Milano.

This modern tonal allegory of time's transcendence represents the passage beyond mortal time and the time of the world in fulfillment of the soteriological promise of Messiaen's Catholic tradition. Accompanied by the hypnotic repetition of the thirty-second note double-dotted eighth note figure, the solo violin's slowly ascending line embodies the gestural figure of an expansive *anabasis*. Yvonne Loriod-Messiaen, Messiaen's wife, explains that her "husband was always struck by Time … [and] often read on the subject of Eternity in the *Summa Theologica* of Saint Thomas Aquinas."[29] For Messiaen, the Apocalypse was the "most enchanting, the most awesome, the most wonderful of fairy tales…. [b]ecause all of it is true."[30] Rebecca Rischin recounts how the "dazzling fairy-tale colors" [31] of the Apocalypse described in Revelation are for Messiaen symbols of the Divine Light. In *Quatuor pour le fin du Temps*, the impalpable tonalities of the "colored emotion"[32] that Messiaen recounts he experienced when as a child he first saw the stained-glass windows of St. Chappelle thus reach their apogee in the ensuing silence of eternity.

The silence to which *Quatuor* ascends symbolically underscores the impossibility of representing the other of time other than by means of allusions to it. In my earlier discussions of music's temporal character, I emphasized how a work's configuration springs from a succession of tones, harmonic complexes, timbral combinations, etc. This configuration, I said then, owes its temporally dynamic character to the operation that draws successive elements together. Both music's temporal features and its affective attributes are thus properties of this schematizing operation. The impossibility of representing something beyond time in an immediately literal way stems from the fact that any representation of a "time beyond time" paradoxically rests on a sound's or a series of sounds' temporal extension. How, we could therefore ask, could music evoke the feeling of "being beyond time" apart from this reliance upon time's rectilinear dimension, which according to the ordinary concept consists of an infinite succession of instantaneous "now" points? Even the representation of silence as an "empty" duration presupposes this duration's temporal extension, as John Cage's account of his compositional technique in *Sonatas and Interludes* forcibly attests.[33] Silence's phenomenal character is indicative of the temporal constraints of representing the alterity of time's other. Similar to the god who in Hölderlin's poem reveals himself as the unknown by way of the upward glance that spans between the earth and the sky, music consequently refigures time's ultimate unrepresentability only through staking out the borderlines between time and the other of time.

Messiaen's *Quatuor pour le fin du Temps* is striking in this regard. The symbolism that Messiaen associates with the work's various movements revolves around the eschatological significance of the biblical passage that inspired the work. Conversely, Messiaen's concern with compositional processes that would annihilate time places the destruction of a sense of forward-directed movement at the center of the work's aesthetic constitution. Messiaen once remarked that time "is one of God's strangest creatures, because it is totally in conflict with his eternal nature, he who is without beginning, without end, without succession."[34] We can hear in this remark the echo of Augustine's effort to wrestle with the question of the being and the nonbeing of time. For Messiaen, as for Augustine, time exists only where there is succession and change; eternity in contrast has no beginning or end. Hence, for Messiaen, whereas "all God's creatures are enclosed in time,"[35] the eternity of God stands outside time.

Messiaen's solution to the problem arising from this opposition between eternity and temporal succession and change sharpens the contrast between Hebraic and Hellenic inspirations regarding the ways that we think about time. For Messiaen, music's metric division into regularly patterned beats gives birth to the time created by music. In contrast, since there is no succession in eternity, for him isorhythmic structures and rhythmic palindromes overturn the traditional system of music's temporal organization through subverting this metric order. Messiaen's "hope for an end to 'the equal and measured time of classical music'"[36] vests the title of his quartet with its double entendre. *Danse de la fureur, pour les sept trompettes* is the most characteristic movement rhythmically in this regard. Together with the unison instruments, which are "made to sound like gongs and trumpets (the first six trumpets of the apocalypse followed by various catastrophes, [and] the trumpet of the seventh angel announcing the consummation of the mystery of God),"[37] added values and augmented, diminished, and retrograde rhythms countermand traditional senses of temporal patterning.

The mix of *Quatuor pour la fin du Temps*' soteriological resonances with the innovative use of modes and rhythmic structures underscores the poetic and technical *tours de force* required when it comes to evoking a sense of eternity. In my reflections on Bach's music, I related the presentiment of abiding grace to an archaic Hebraic inspiration, according to which eternity is irreducible to an immutable present. Replacing the metaphor of the "arrow of time" with that of "cyclical time," I argued then, voids the significance of this presentiment with respect both to the liturgical value of Bach's music and to time's deficiency in relation to the plenitude of eternity. Representations of something beyond time that obliterate movement and change are more Hellenic than Hebraic in this respect. The Platonic account of the revolution of the celestial bodies as the moving image of eternity's immutability is a case in point, as it binds the cosmological concept of time to movement and change. The paradox that music evokes a sense of eternity through resounding in time highlights the enigma of invoking the alterity of an order that escapes our power to represent or master it through abolishing processes of movement and change. Consequently, where Greek inspirations regarding thought about time prevail, representations of time's annihilation or destruction seemingly draw near the myth of eternal return.

The desire for light and for jubilant singing that in his prefatory note to *Abîme des oiseaux* Messiaen contrasts with the abyss of time sets the hope of transcendence against this myth of eternal return.[38] This abyss is the emblem of time's ontological deficiency. Time, with its sorrows and weariness, wears away all things that are subject to its devouring power. Intimations and signs of time's deficiency vis-à-vis the fullness of eternity consequently bear on the manner of our consent to an existence that is given to us.[39] Setting time's deficiency against eternity is thus a testament to a center of alterity announced by the superlatives of the other of time's exteriority, anteriority, and height. From this point of view, the bird songs that in *Abîme des oiseaux* symbolize the desire for flight also sound out the joy that we have in our part in being. Where, in this movement, the solo clarinet's pensive meditation reverberates with the sadness of our finite condition, birds

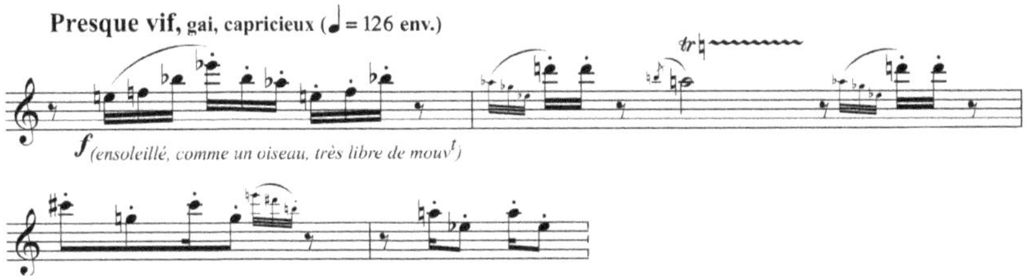

Example 6.2 Olivier Messiaen, *Abîme des oiseaux*, mm. 13–16.

Source: *Quatuor pour la fin du temps*. Music by Olivier Messiaen.

are emblems of the opposite of Time. Bird songs are akin to the time of the angels in this respect. For angelic time "is the *aevum*, that is 'the intermediary between time and eternity.'"[40] This angelic time is neither the eternity of God nor the fragile temporal order of human affairs. Ernst Bloch reminds us that according to Thomas Aquinas, angels "are not contained in space like a body, but they themselves contain their space, and they enclose it instead of being enclosed by it."[41] Like angels, bird songs are free of the cares and constraints of this temporal world. Spirited flourishes offer a presentiment of this carefree time in imitation of the blackbird's song (Example 6.2).[42] Rising above the abyss, this insouciant song—perhaps like the praise of angels—rejoices in taking flight from time's burdensome weight.

Contra-finality

The circumstances surrounding *Quatuor pour la fin du Temps*' creation clearly invite speculations into the connections between its apocalyptic inspiration and the political cataclysm into which the world had been thrown. Although Messiaen rejected the idea that the work's theme was specifically related to his internment at Stalag VIII A, the moral conflagration and political madness of fascist aggression might with some justification be compared to the biblical prophecy of the end times. In 1947, the initial countdown to global catastrophe on the Doomsday Clock was set at 7 minutes. In 2017, the Doomsday Clock was advanced to 2.5 minutes until midnight. The march of history, it seems, has brought us only closer to total annihilation. We can scarcely believe that the future promises to bring better times, or that the project of the Enlightenment has succeeded in advancing the causes of peace and justice. In view of the political disasters and historical catastrophes that continue to pile up, the end sometimes appears to be all but inevitable.

The war that Jean-François Lyotard insists that we wage on totality in answer to a modernist nostalgia for the whole presages a different destiny for the end of time than does Messiaen's *Quatuor pour la fin du Temps*, which draws its inspiration from the Apocalypse's soteriological message. Lyotard's seminal diagnosis of the postmodern condition vigorously denounces the violence and terror that he identifies with Enlightenment thought. For him, the price that we have paid for the reconciliation of concepts and sensible appearances (in accordance with Kant's thought) or the totalization of time through elevating the thought of history as a whole to the level of the absolute (in accordance with Hegel's philosophy) is reason enough to renounce modern aesthetics and its nostalgic allegiances. On Lyotard's analysis, modern aesthetics falls short of the sentiment of the sublime by offering the "reader or viewer matter for solace or pleasure"[43] despite modern art's allusions to a surplus or excess that defies the imagination's power to provide an adequate representation of the sublime's illimitable magnitude or dynamic power. In contrast, the postmodern "puts forward the unpresentable in presentation itself"[44] in order to more vehemently impart a sense of this surplus or excess. As witness to the unpresentable, the postmodern thus vacates the modernist demand for a totalizing reconciliation.

By emphasizing the distinction drawn by Lyotard between modern and postmodern aesthetics, I want to take up again the question of the feeling of the sublime's significance with respect to music's explorations of the difference between the being and the nonbeing of time. In my discussion of Bach, I indicated how the feeling of the sublime in his music marks out a threshold for religious aesthetics. Contrary to metaphysical interpretations, I argued that the feeling of height in Bach's music countermands the temptation to lay hold of the absolute. I therefore also cautioned against confusing moral injunctions that accompany the assignment of responsibility with aesthetic presentiments. The proximity of the feeling of the sublime with the glory of the infinite gives rise to a temptation to vest music's mimetic explorations of ways of inhering in the world with a religious or moral fervor. In the final chapter, I will address at some length the distinction between moral and political injunctions and the truth to which works lay claim within the broader context of a reflection on aesthetic experience's lateral transposition onto the planes of ethics and politics. My more immediate interest in this chapter is in the way that the feeling of the sublime provides a philosophical touchstone for interrogating how music revalues time through evoking a sense of being beyond time. Lyotard's analysis of the postmodern sentiment of the sublime accordingly plays a part in the following discussion. However, in view of my larger concern with the ways that music refigures the difference between time and the other of time through staking out the boundaries between them, I also want to take into account how a hermeneutical riposte to the loss of credibility of "grand" historical narratives opens a different avenue of thought regarding history's temporalization in the light of music's simulacra of the end of time.

The transgression of traditional narrative and musical conventions in Philip Glass's *Einstein on the Beach* (1976) exemplifies the kind of radical departure from modern aesthetics that for Lyotard signals the sempiternal advent of the *post-modern*, which according to him would then "have to be understood according to the paradox of the future (*post*) anterior (*modo*)."[45] Developed in collaboration with Robert Wilson, this "portrait opera" celebrates the "almost saintlike originator of the theory of relativity"[46] who also has been widely identified with the beginning of our catastrophic nuclear age. Glass explains that the story to be drawn from intercut sequences of recurring scenes—Train scenes, Trial scenes, and Field/Spaceship scenes, along with interspersed "portrait-scale Knee Plays"—was left entirely to the audience's imagination. On Glass's account, the poetic vision that he and Wilson constructed rests on the quasi-development of recurring images. Altering an image each time it reappears so that this image becomes "more abstract and, oddly enough, more powerful"[47] plays a major role as a dramaturgical device. Contrasting slow trial scenes with fast dance scenes, for example, structures the flow of the work's visual and musical processes without constituting a narrative development. The work, which begins by depicting a nineteenth-century train and ends with taking leave of a worldly bounded existence in a spaceship, accordingly trades on the juxtapositions of altered recurring images for its theatrically dramatic effect.

This resolutely anti-narrative "portrait opera" breaks emphatically with traditional plot conventions. By concluding episodes without imposing any meaningful order on a scene or an act, finales and epilogues supplant the "sense of an ending" that in traditional narratives draws an otherwise endless series of incidents to a close. In traditional narratives, a story's configuration as a whole turns on the end point in which this otherwise endless succession of individual incidents terminates. Hence, the significance of each incident rests on the way that each contributes to the development of the plot. Each incident is a necessary part of the story only by reason of the way that it advances the plot toward its concluding point. Correlatively, the plot transforms the contingencies of its separate episodes into necessary elements of the story as a whole. The structural function of narrative closure is readily discernable in the act of retelling rather than telling a story, as Ricoeur points out. By recapitulating the "initial course of action in its terminal consequences," [48] we can invert time's supposedly natural order. "[R]eading the ending in the beginning and the beginning in the ending"[49] also underscores how a story's temporal configuration as a whole is the ground of the teleological thrust imputed to it. In sum, the end in which the actions recounted terminate is a narrative artifice that encloses the world populated by the story's character in its own universe.

The express absence in *Einstein on the Beach* of any "direct connotative 'meaning'"[50] clearly disrupts traditional expectations and conventions. Shifting images in contrasting sections subvert any sense of narrative coherence at the same time that they invite an interpretation based on the way that the succession of shifting images coheres. According to Glass, audience members supply a meaning by personalizing their experiences of the opera so that any story attributed to this otherwise resolutely abstract work is a product of her or his imagination. This apparent need to supply a meaning by drawing a story from the sequences of staged images and the accompanying music puts into play the imagination's operative role. In short, the audience members grasp the succession of scenic elements as meaningfully related by schematizing the way that these successive elements contribute to the development of an intelligible whole. The audience members consequently activate the implicit contract between an author and her readers and between a composer and her listeners by undertaking the work of drawing a configuration from the succession of images and sounds.

Hence, if by breaking with the real time of our everyday lives fiction fulfills its own law, the dissolution of our every expectation regarding a plot (in the case of narrative) is not necessarily the final word. Just as the work's retreat from the practical order of our everyday experiences is the condition of its subversive power, music (like narrative) occasions new or different experiences only through its temporal configuration. Simulacra of the "end of time" are pointedly fictitious in this respect. The work's closure within the world projected by the work is consequently the condition and the spring of the work's ontological vehemence and force.

The crisis of closure and its concomitant, the wearing away of the narrative paradigm in which concordance wins out over discordance, are comparable to representations of the "end of time" that trade on the apocalyptic model's qualitative transformation in this regard. Ricoeur stresses that to the degree that predictions regarding the world's end have been invalidated, the imagery of the Apocalypse has shifted "from the last days, the days of terror, of decadence, of renovation, to become a myth of crisis."[51] Hence, where once the world's end was imminent, the apocalyptic model's conversion into a myth of crisis transforms an impending end into one that is immanent within the forces and processes driving history along its disastrous course. In view of this crisis, we might wonder what a culture would be in which no one knew what it meant to recount one's actions or to tell one's story. Furthermore, we cannot ignore the possibility that the style of historical experience that has dominated in the West is dying out.[52] Hence, we might also wonder whether we have reached a point where we no longer have any experiences to share and whether the anti-teleological thrust of representations of the end of time is the emblem of a desire for a return to a time *in illo tempore*.

Through converting tonal closure's paradigmatic function into a means of representing the end of time's contra-finality, *Einstein on the Beach*'s "apocalyptic grand finale"[53] evinces the crisis of closure (which is symptomatic of the time of crisis to which I previously referred) by upending tonality's ontotheological underpinnings. Through transfixing the "time-honored closing"[54] that vests dissonance's resolution with its normative force, the recurring "cadential" formula transforms the traditional sense of closure into an endless ending. Resolving on a harmony that is a half step lower than the one on which it begins, the cyclic repetition of this recurring "cadential" pattern (f-D♭-B♭♭/A-B-E) voids the impulse that in traditional tonal music fuels the drive toward resolution (Example 6.3). The recurrence of this standing "cadential" pattern accordingly consigns memory and anticipation (the "usual psychological devices of programmatic music whether Baroque, Classical, Romantic or Modernistic"[55]) to oblivion. This sempiternal ending's immemorial beating in the Spaceship scene bids a final farewell to all temporally earth-bound existence. It is as if, in answer to the catastrophic events of our nuclear age, *Einstein*'s apocalyptic finale recalls the archaic rhythm of an eternal return that heralds the passage beyond.

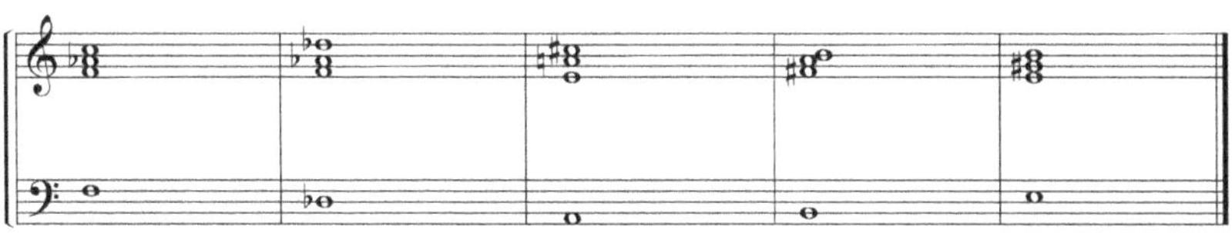

Example 6.3 Recurring "cadential formula".

Hermeneutics and the feeling of the sublime

The divinely intemporal order that Lyotard locates at the heart of the real sentiment of the sublime reverberates in this postmodern simulacrum of the *nunc stans*. For Lyotard, the grace to which all genuine music aspires escapes our power to recount events and to draw a meaning from them by gathering them together. Accordingly, all genuine music seeks to exempt itself from the temporal condition in which successive incidents contribute to the formation of expressive figures as they unfold. The explosion and dispersion of the divine monad, following which we perceive sound through its temporal extension, opens the possibility for synthesis in giving birth to time.[56] In contrast, the music of the spheres *qua* divine monad is as a single chord that neither begins nor ends and hence is neither limited, nor constrained, by the synthetic operation that is the hallmark of the imagination's schematizing power.

The real sublime sentiment according to Lyotard therefore militates against reason's demand for totality by stripping the mind of its power for synthesis. In Chapter 3, I outlined how for Kant the feeling of displeasure that he attributes to the imagination's inability to provide an adequate representation of the object instills in us a feeling of respect for our supersensible vocation. For Kant, the feeling of the sublime consequently harmonizes the imagination's power of presentation with reason's moral ideas through giving rise to the thought of the unconditioned (to which it also attunes us) as the measure of reason's autonomous law.[57] This demand for the unconditioned, I said then, would be put to the test in this chapter when I set it against the war that Lyotard maintains must be waged against totality. If, by emphasizing how the contra-finality of an ending without end evinces the crisis of a sempiternal present I relate this representation of the "end of time" to the feeling of the sublime, it is in order to set out the hermeneutical limits of the demand that in one way or another motivates both Kant's and Lyotard's analyses of this feeling.

For Lyotard, the real sentiment of the sublime irrupts from the mutual destruction of moral universality and aesthetic universalization in the "violence of their differend."[58] Hence, unlike Kant, Lyotard refuses to harmonize the feeling of displeasure evoked by the imagination's inability to provide an adequate representation of the sublime object with supersensible ideas of reason. The differend, Lyotard insists, "cannot demand, even subjectively, to be communicated to all thought,"[59] since the differend is the trace of a *presence* that exceeds the imagination's capacity to give it a sensible form. For Lyotard, the excess that Kant deems to be repulsive to our sensible intuition can be felt only at the edge of the break between sensible presentation and the totality demanded by reason.[60] Accordingly, the transcendental supplement that he maintains rescues Kant's analytic of the sublime by means of the Idea of reason is for him the sign that the infinite—which according to Kant elicits the feeling of the sublime—can be revealed only through its negative presentation.

By distinguishing between a mode of repetition that gives rise to a metaphysics of ideas and a mode of repetition that he maintains is aesthetic, Lyotard consequently sets an "ontology of being as non-being"[61] against the moral impulse that Kant identifies with the sublime. Where for Kant the violence done to the imagination in striving to provide a sensible intuition of the sublime object awakens us to the idea of totality in accordance with reason's demand (see Chapter 3), for Lyotard the feeling of terror denies the solace that reason derives by adding the thought of totality to a form's unboundedness.[62] For Lyotard, the mix of fear and exaltation in the feeling of the sublime is irreducible to any moral sentiment.[63] Hence, unlike the beautiful, which offers an assurance of the unity of a transcendental subject, the sublime blazes up suddenly without the promise of any future as the desire for the illimitable that thinking feels in its conflicted state of happiness and unhappiness.

While for Lyotard this conflicted state of happiness and unhappiness places Kant's transcendental subject under erasure, the disproportion between reason's demand for totality and the imagination's power to present the sublime object in a sensible form is more fundamentally an index of our affective fragility. When in Chapter 3 I related joy and sorrow to the good that we desire and the evil that we flee, I also emphasized the importance of the distinction between pleasure and happiness. Pleasures place their stamp on our bodily experiences; happiness transcends these partial pleasures' finite perfection through aiming at the good that we desire to be. If the totality demanded by reason has the same breadth as happiness, as

Ricoeur maintains, it is only insofar as feelings of happiness are signs of as yet unfulfilled orders of freedom in the kingdoms of love of others and of the devotion to exemplary ideals. Consequently, reason's specific desire is Eros inasmuch as through Eros we affirm the part that we have in being.

The totality demanded by reason undoubtedly marks a limit with regard to our ability to think the unconditioned. When in the next chapter I relate the "rule" summoned by exemplary moral and political acts to the exercise of judgment, I will oppose the power to act in response to the exigencies of the situations to the idea that the fulfillment of this demand can be posited in advance. Like individual works, exemplary acts shatter congealed conventions and habits of thought. Ricoeur's post-Heideggerian return to Kant through Hegel is thus motivated in part by the way that Kant's style of philosophizing reserves a place for the logic of hope. Hegel's system is "one that is written from the end toward the beginning,"[64] so that its rational character is a function of a style of thinking that dares to elevate itself to the level of the absolute. In contrast, Kant's philosophy underscores the limits of our capacities and powers. Even so, Ricoeur is critical of the subtle tutelage exercised over the ethical sphere by Kant's first *Critique*'s transcendental procedure. Hence, for Ricoeur, we can enter into the ethical problematic only via experiences of passivity and the attendant feelings of being-affected before taking these feelings up in a reflection on our fragility and fallibility.[65]

That the feeling of the sublime raises the question of reason's demand for totality at all underscores how the imagination's failure to provide a sensible intuition for an unbounded object redounds on the subject's self-constitution. This inability on the part of imagination to provide an adequate representation for the thought of the absolute bears the stamp of the pathos that inheres in the feeling of the sublime. Born from the difference between the finite and the infinite, this pathos is the emblem of the deficiency that strikes at the heart of our experiences of time. When Lyotard insists that the pure matter of sound only reaches the subject "at the cost of surpassing, or 'sub-passing', its capacity for synthetic activity,"[66] the imagination's failure becomes the sign of the subject's extinction. Deprived of the imagination's schematizing power, there is no subject that could refer to itself. This, Lyotard remarks, "is how Epicurus circumscribes death: if it's there, I'm not there; so long as I'm there, it's not there."[67] The *nuance* that gives the ear over to the play of incomparable differences in repetitive forms of music is accordingly a hallmark of the sublime's postmodern resonances.[68] For Lyotard, the time of the atom and the time of the god are neither temporal nor temporalizing but are instead the boundary posts that mark the limits of thought about time. The negative judgment regarding the question of the subject's possibility "constituted according to the model of the *Ich denke* and the temporality required for theoretical thinking"[69] consequently lays bare the abyss over which Nietzsche recommended casting an Apollonian veil. The question thus arises: If the advent of the order of the inhuman is not a return to a time *in illo tempore*, does this negative judgment regarding the possibility of the subject supplant the mythic desire for cosmos with a fascination for chaos in pursuit of its libidinal excesses?

Specters of the "end of time"

The question raised by the preceding consideration of the sublime leads again to my thesis that music revalues time through exploring the boundaries between time and the other of time. "Eternity" experiences, I have said, refigure time's ultimate unrepresentability through staking out the borderlines between time and the other of time. However formless they may be, the feelings (*Stimmungen*) evoked schematize affective dimensions of our experiences in response to time's inscrutable character. Music's affective force revolves around its power to affect us through modulating moods and feelings that we make our own. The metamorphosis in music of everyday, bodily emotions thus opens a vast field for experimenting with a range of affective possibilities.[70] Feelings evoked in response to the disproportion between time and its other upend the sense of harmony that Kant identifies with the free play of imagination and understanding in aesthetic judgments of beauty. Yet, through foregrounding the imagination's failure to provide an adequate representation for the illimitable, the infinite, or the unformed, these feelings also bear on the self's constitution. Feelings *as felt* on the illimitable or the infinite affect the self in response to the alterity of an order that exceeds the subject's grasp. Feelings belonging to the order of the sublime thus evince how music's evocation of its affective tonalities revalues time in accordance with the representation of something beyond it. Such feelings accordingly sound out regions of alterity that surpass our capacity to grasp or master them in answer to the deficiency born from the difference between the being and the nonbeing of time.

The plurality of these regions of alterity gives us a good indication of the different ways in which time's other constitutes a limit both with regard to thought about time and with respect to time's revaluation vis-à-vis its ultimately inscrutable character. The original nothingness that in Augustine's reflections is anterior to the universe's creation *ex nihilo* is not the pure nothingness veiled for fear that we would die if we dared to contemplate it. This original nothingness's anteriority, superiority, and height also differs from the final deliverance from the "eternal return to existence by force of *karma*, the law of universal causality,"[71] through abolishing and transcending *karma*'s cosmic power. The "tragedy of the sempiternal"[72] that in *King Lear* operates within the order of the protagonists' and antagonists' misfortunes is especially instructive in this regard. Since "Lear's torment tends toward a continually postponed conclusion.... [,] the end ... is only an image of the horror of the time of crisis."[73] This quasi-eternal time of crisis thus marks the conversion of the apocalyptic model's imminent end into an immanent end that is seemingly infinitely postponed (see above). But then, as Ricoeur pointedly asks, "What is an immanent end when the end is no longer an ending?"[74] Hence, we could also ask in turn if the contra-finality of the "end of time" signals the advent of a sempiternal crisis through usurping the place of the other of time.

Does the idea of "pure, punctual presence"[75] that Lyotard identifies with the subject's extinction similarly convert the time of crisis into the crisis of time? Lyotard admits that this idea is extremely problematic, since "pure, punctual presence" cannot be conceived, experienced, or felt in accordance with the forms of sensibility available to us. Rather as an idea, "pure, punctual presence" is a function of the subreption of the subject's capacity to grasp together a manifold in a single intuition. The idea that for Lyotard is the emblem of repetitive music's evocation of the real sentiment of the sublime has an exemplar in music that seemingly embodies itself as a "pure sound-event."[76] Yet even this "event" extends over sometimes long temporal expanses. Jacques Attali consequently argues that like hyperrealism, repetitive music advertises itself as the "murder of creativity" [77] through laying waste to time. The form of political economy with which Attali identifies repetitive music liquidates the relations among the human beings that it governs through fetishizing the circulation of differences. Hence, Attali's denouncement: The "stockpiling of use-time in the commodity object is fundamentally a herald of death."[78]

Situating repetitive music within the context of a political economy that produces the demands that it meets, however, cannot fully account for repetitive music's cultural and historical significance. On the contrary, the sense of the sempiternal evoked by the conversion of music's temporal extension into the simulacrum of the *nunc stans* is indicative of the crisis into which the present is thrown when, torn between two fleeing horizons (past and future), the force that the present has as the time of action is arrested. Time's revaluation by this simulacrum of a "standing present" is unique in this respect. Neither the ontological deficiency that Augustine's meditations bring to light by setting time against eternity nor the ritual fault evinced by the difference with respect to myth between cosmos and chaos comes close to usurping the force of the present.[79] Hence, we might wonder whether the war that Lyotard insists must be waged on totality marks a retreat from the burden and responsibility of preserving the tension between the horizons of a past that has been surpassed and of as yet unfulfilled demands and expectations. Could it be that, in fleeing the terror of history that Eliade attributes to the loss of religious attitudes in the modern world, the simulacrum of "pure, punctual presence" supplants the ritual return to the *illud tempus* that transcends the nothingness of death through setting cosmic time against this terror? The form of repetition that Lyotard identifies with the aesthetic takes the place of the nothingness of death. Since in this mode of repetition there is no subject that could refer to itself or identify itself as such, this aesthetic mode of repetition extinguishes death in accordance with the order of the inhuman.

The representation of the temporal ruins of Akhnaten's world in the Epilogue of Glass's period opera invokes this beatific specter of the end of time. According to Glass, Akhnaten proclaimed a new monotheistic religion based on Aten (the sun). Paul John Frandsen remarks that *Akhnaten* offers little real insight into the ancient Egyptian world it portrays.[80] Notwithstanding, this prismatic palimpsest's poetic representations of two distant worlds register the anxiety over death that for Glass any portrayal of Ancient Egypt must include. The depiction of the ruins of Akhnaten's "now-dead city"[81] in the Epilogue stands as a testament to time's corrosive power. Yet in setting these present-day ruins in relief, the Epilogue also takes on "a lyrical, almost distant quality, as if death and the passage of centuries … have softened the violence of those moments, making them almost beautiful."[82] Following the restatement of the opera's main materials (Trilogy theme, the music associated with Akhnaten's downfall, music accompanying scenes of violence and destruction, and a "passing reference to Akhnaten's Hymn from Act II"[83]), a final "ending" motif draws the opera to a close.[84] As the Epilogue concludes, the recurring motif of repeating A's and E's, together with a final appearance of the Trilogy theme, casts a shroud over the silent ruins of a now desolate landscape.

The beatific aura that envelops this sublimely intemporal scene echoes *Einstein on the Beach's* transvaluation of the traditional cadential ending. If as Pelikan suggests an era's interpretation of death is indicative of its true spirit, these simulacra of the end of time register the meaning of the present in accordance with the destruction of reason's demand. Death, Glass stresses, was "the great subject of ancient Egyptian culture."[85] Like Bloch, who sees the pyramids as timeless monuments of a sacred geometry that entombs a "space without time,"[86] Glass regards the pyramids' size and grandeur as an attempt to overcome death in the eternity of the silent sepulcher. For Bloch, stitching up the dead as a dry mummy fulfilled the desire to become like stone that shaped the utopian "landscape of Egyptian art."[87] The timeless rigidity of the Egyptian kingdom of the dead thus perfected earthly existence by prolonging this kingdom forever. As the distant reference point for portraying the splendor of this ancient world's temporal ruins, the sacred geometry that in this death-conscious culture extinguishes time thus also becomes the emblem of the eternal present's usurpation.

Nietzsche's exhortation—"Sing! speak no more!"[88]—is indicative of the tragedy of the sempiternal that follows in the wake of the destruction of reason's demand for totality and of the ensuing reversal of the Hegelian temptation to grasp the oneness of time in the eternal present. In proposing his own interpretation of the state of despair that Nietzsche experienced while writing *Thus Spoke Zarathustra*, Gadamer sets the will to life against the wisdom of self-consciousness and its knowledge of the inevitability of death. This knowledge "confronts itself in the eternal recurrence of the same just as much as the will to eternity does."[89] Because this conflict cannot be resolved through *logos*, the exhortation to sing offers the only hope for a victory over death, which for Gadamer can be overcome only through accepting life's eternal recurrence. Nietzsche's pronouncement thus affirms the "desire for the eternity of life."[90] Jacque Derrida stresses that time "cannot be conceived on the basis of a present and the self-presence of a present being."[91] The difference that fuels the destruction of metaphysical conceits thus also bears the mark of the alterity of an order that is antecedent to time's existential constitution.

Whether this difference between time and the alterity of an order that exceeds our grasp is the sign of an original nothingness anterior to the creation of the cosmos, or whether this difference motivates the desire for the eternal return of the same, music's affective tonalities register the deficiency that places its stamp on our temporal finitude. The feeling of the sublime that promotes the subject's extinction is no doubt indicative of a condition resulting from the war waged by postmodernism on totality. Nevertheless, this condition's historic irruption blocks the way to a retreat into myth. In the next chapter, I therefore intend to explore how aesthetic experience's lateral transposition onto the planes of ethics and politics augments my reflections on music's mimetic revaluations of time.

Notes

1 Mircea Eliade, *The Sacred and the Profane: The Nature of Religion*, trans. Willard R. Trask (Orlando, FL: Harcourt, 1959), 70; see chapter 2.
2 Eliade, *The Sacred and the Profane*, 108; see Paul Ricoeur, *Figuring the Sacred: Religion, Narrative, and Imagination*, trans. David Pellauer, ed. Mark I. Wallace (Minneapolis, MN: Fortress Press, 1995): "[M]yth as recited and inserted in the ritual of renewal makes *homo religious* participate in the efficacy of the sacred" (51).
3 Eliade, *The Sacred and the Profane*, 71.
4 Mircea Eliade, *The Myth of the Eternal Return: Or, Cosmos and History*, trans. Willard R. Trask (Princeton, NJ: Princeton University Press, 1954), 150.
5 Eliade, *The Sacred and the Profane*, 129.
6 Martin Heidegger, *Poetry, Language, Thought*, trans. Albert Hofstadter (New York: Harper and Row, 1971), 227.
7 Heidegger, *Poetry*, 226. By disclosing that which conceals itself, the sky reveals the unknown god not by enabling us to "wrest what is concealed out of its concealedness, but only by guarding the concealed in its self-concealment. Thus the unknown god appears as the unknown by way of the sky's manifestation" (223).
8 Heidegger, *Poetry*, 226.
9 Heidegger, *Poetry*, 223; original emphasis.
10 Mircea Eliade, *Myths, Dreams, and Mysteries: The Encounter between Contemporary Faiths and Archaic Realities,* trans. Philip Mairet (New York: Harper and Row, 1960), 24; original emphases.
11 Ricoeur, *Figuring the Sacred*, 49. The experience of the sacred does not

> inscribe itself within the categories of the Logos of proclamation and its transmission and interpretation. The "numinous" element of the sacred is not first of all something to do with language, even if it may become so to a certain extent subsequently. To speak of "power" is to speak of something other than "speech," even if the power of speaking is thereby implied. It is a power that does not pass over completely into articulation since it is the experience of efficacy par excellence.

The sacred can manifest itself in objects that the believer venerates such as a geographical place, a body of water, a rock, or a tree, for example. Hence, the sacred is in the "same position in relation to its manifestation as the ideas of our reason are in Kant in relation to their presentation in the products of the imagination" (50). See also my discussion of the sublime in Chapter 3.

12 Ricoeur, *Figuring the Sacred*, 50.

13 Gilbert Rouget, *Music and Trance: A Theory of the Relation between Music and Possession*, trans. Brunhilde Biebuyck (Chicago, IL: University of Chicago Press, 1985), 121:

> In the dimension of time, music modifies our consciousness of being to an even greater extent. It is an architecture in time. It gives time a density different from its everyday density. It lends it a materiality it does not ordinarily have and that is of another order. It indicates that something is happening in the here and now; that time is being occupied by an action being performed, or that a certain state rules over the beings present.

14 Rouget, *Music and Trance*, 28. For Rouget, "nonidentificatory trance inspiration or communion trance" that seem to be characteristic of Islam, Judaism, and Christianity distinguish these forms of trance from those that depend on the ritual identification with a deity. According to him,

> the autochthonous theory is that possession is produced by the will of a divinity who, for one reason or another, wishes to incarnate him- or herself within a particular individual. The god is the one who selects the person in whom he wishes to incarnate and not the reverse: a person does not select a god for whom he then becomes the vessel. On the surface at any rate, the relation as experienced by the possessed person is passive rather than actively willed.
>
> (34)

15 Rouget, *Music and Trance*, 122; see Judith Becker, *Deep Listeners: Music, Emotion, and Trancing* (Bloomington, IN: Indiana University Press, 2004). Becker argues that trancing attests "to the divine presence in one's midst, legitimizing the religious beliefs and practices of the community" (1). For her, music

> provides a link between alternate selves and alternate places and alternate times that become real places and real times in trance experiences.... One is moved from the mundane to the supra-normal: another realm, another time, with other kinds of knowing.
>
> (27)

16 John Blacking, *How Musical Is Man?* (Seattle, WA: University of Washington Press, 1973), 28; see my "Being, Transcendence and the Ontology of Music," *The World of Music* 51, no. 2 (2009).

17 Blacking, *How Musical Is Man?*, 48. By distinguishing between music's social function and its aesthetic value, Blacking in effect identifies music's power to create superreal "worlds of time" with its ontological vehemence. See my "Being, Transcendence and the Ontology of Music."

18 Ali Jihad Racy, *Making Music in the Arab World: The Culture and Artistry of Ṭarab* (Cambridge: Cambridge University Press, 2003), 6. Racy explains that *ál-fann ishās* ... means 'art [namely music], is feeling'" (4). *Ṭarab* has been associated with "either joy or sadness, or sometimes with a combination of both. In this case, the two emotions represent opposite but interconnected poles, and as such can be said to belong to the same experiential complex" (203–204).

19 Racy, *Making Music*, 120:

> in such a state, the performer finds himself captivated by the mode, particularly the intervallic and tonal components. He feels haunted by the

> tonic pitch and the intervallic structure, but is also fully prepared to evoke the powerful ṭarab effect of the mode. Thus the strongly felt presence of an established tonic and related intervals and notes of emphasis makes it possible to view salṭanah as a form of modal, and by implication tonal and intervallic fixation, or essentially as modal ecstasy.
>
> (120–121)

Racy remarks that Rouget "stressed that of all world peoples, Arabs make the strongest association between music and trance and that such associations applied to both sacred and secular practices" (3; see 199). See also Rouget, *Music and Trance*, 255.

20 Eliade, *The Myth of Eternal Return*, 153–154.

21 Eliade, *The Myth of Eternal Return*, 161. Hegel's philosophy of history

> still preserves something of the Judeo-Christian conception: for Hegel the historical event was the manifestation of the Universal Spirit.... Thus, in Hegel's view, the destiny of a people still preserved a transhistorical significance, because all history revealed a new and more perfect manifestation of the Universal Spirit.
>
> (148–149)

See Eliade, *The Sacred and the Profane*, 111–112.

22 Eliade, *The Myth of Eternal Return*, 160.

23 *Quatuor pour la fin du Temps* was premiered by Henri Akoka (clarinet), Jean Le Boulaire (violin), Étienne Pasquier (cello) and Olivier Messiaen (piano). On Messiaen's account, over 5,000 prisoners attended. Messiaen insisted that the work's theme was not specifically related to his imprisonment. See Rebecca Rischin, *For the End of Time: The Story of the Messiaen Quartet* (Ithaca, NY: Cornell University Press, 2003), 30 ff. Étienne Pasquier recounts that during one of the rehearsals, a German officer told Messiaen: "In a few weeks, there will be a return of prisoners to Paris. Don't miss the train." (Étienne Pasquier, quoted in Rischin, 72). Messiaen turned to Pasquier and asked: "How can I return?" The officer replied, "Because you're *soldat musiciens* [who were drafted to serve as musicians]. You were caught unarmed" (Pasquier, quoted in Rischin, 73). Pasquier points out that he had a rifle when he was captured. Messiaen, Pasquier, Boulaire, and Akoka were released about a year after.

24 See Rischin, *For the End of Time*, 51. In Rischin's view, Messiaen elevated the statement "*Il n'y a plus de Temps*" (which she points out appears in many translations of the New Testament as "*Il n'y aura plus de délai*"), to the sublime. Messiaen explained that, "There are people who understand 'there will be no more delay.' That's not it. [It's] 'there will be no more Time' with a capital 'T'" (quoted in Rischin, 51). Messiaen's conviction that one rejoins eternity through leaving the human dimension of our temporally finite existence seems to support an interpretation that identifies eternity with the cessation of time as measured by the revolution of the heavenly bodies. Hence, for him, the end of time for which the quartet was written signals the "end of the notion of past and future" (Messiaen, quoted in Rischin, 64). Some commentators interpret the prophecy to mean that there will be no more delay of the coming of Christ's kingdom when the seventh trumpet sounds. See for example "Revelation 10: Gill's Exposition of the Entire Bible," Bible Hub, accessed November 11, 2016, http://biblehub.com/commentaries/gill/revelation/10.htm. See also Richard D. E. Burton, *Olivier Messiaen: Texts, Contexts, and Intertext (1937–1948)*, ed. Roger Nicholas (New York: Oxford University Press, 2016), 45–46; Jeremy Begbie, *Resounding Truth: Christian Wisdom in the World of Music* (Grand Rapids, MI: Baker Academic, 2007), 174.

25 Père Jean-Rodolphe Kars, "The Works of Olivier Messiaen and the Catholic Liturgy," in *Olivier Messiaen: Music, Art and Literature*, ed. Christopher Dingle and Nigel Simeone (Burlington, VT: Ashgate, 2007), 331.

26 Olivier Messiaen, *Quatuor pour la fin du Temps: pour violon, clarinette en si bémol, violoncelle et piano* (Paris: Durand Éditions Musicales, 1942), i: "*Sept est le nombre parfait, la création de 6 jours sanctifée par le sabbat divin….*" For Messiaen, the number seven, which is the sum of the perfect mathematical number, four, and the number of the Holy Trinity, is also the symbol of Christ as the union of heaven and earth.

27 Messiaen, *Quatuor pour la fin du Temps*, i: "*le 7 de ce repos se prolonge dans l'éternité et devient le 8 de la lumière indéfectible, d l'inaltérable paix.*"

28 Messiaen, *Quatuor pour la fin du Temps*, ii: "*Sa lente montée vers l'extrême-aigu, c'est l'ascension de l'homme vers son Dieu, de l'enfant de Dieu vers son Père, de la créature divinisée vers le Paradis.*"

29 Yvonne Loriod-Messiaen, quoted in Rischin, *For the End of Time*, 126. Loriod-Messiaen comments:

> In the Apocalypse there was this moment called Eternity in which the universe would be no more. So when Messiaen read the Apocalypse, he would get the desire to look for new rhythms. Precisely the nonretrogradable rhythms that you have in the Quartet.

See Andrew Shenton, "Observations on time in Olivier Messiaen's *Traité*," in *Olivier Messiaen: Music, Art and Literature*, ed. Christopher Dingle and Nigel Simeone (Burlington, VT: Ashgate, 2007). According to Shenton, "Messiaen's aspiration to eternity is a way of expressing his hope of everlasting life and is directly related to the discussion of God's eternity with which he begins the *Traité*" (175). See also Andrew Shenton, "Five Quartets: The Search for the Still Point of the Turning World in the War Quartets of T. S. Eliot and Olivier Messiaen," in *Messiaen the Theologian*, ed. Andrew Shenton (Burlington, VT: Ashgate, 2010), 154–157; Anthony Pople, *Messiaen: Quatuor pour la fin du Temps* (Cambridge: Cambridge University Press, 1998). Pople points out that the section from the *Diptyque* transcribed in *Louange à l'Immortalité de Jésus* is the "'essay on … blessed eternity'" (81). In *Quatuor pour la fin du Temps* the music is "transposed up a major third from C major to E major" (81). Pople comments that among the few discrepancies between the two works, the substitution in *Quatuor* of a "hypnotic … [sixteenth—double dotted eighth note] rhythm in the piano part for the sustained notes of the original organ writing" stands out.

30 Loriod-Messiaen, quoted in Rischin, *For the End of Time*, 126.

31 Messiaen, quoted in Rischin, *For the End of Time*, 56; from Olivier Messiaen, "Address Delivered at the Conferring of the Praemium Erasmianum on 25 June 1971, in Amsterdam," in *Contributions to the Spiritual World of Olivier Messiaen: With Original Texts by the Composer*, Comp. Almut Rössler, trans. Barbara Dagg and Nancy Poland (Duisburg: Gilles and Francke, 1986), 44.

32 Rischin, *For the End of Time*, 56. Rischin claims that *Quatuor pour la fin du Temps*

> is unique among Messiaen's compositions in that its "sound-painting" stems not merely from the composer's life-long fascination with color, but from his physical deprivation as a prisoner in Stalag VIII A. In his commentary on the seventh movement in the preface, Messiaen recalled in dazzling poetic prose the colored dreams that inspired him to reread Revelation.
>
> (56–57)

> In my dreams, I hear and see classified chords and melodies, common colors and forms; then, after this transitory stage, I pass into unreality and lose

myself in a rapture to a whirling, a gyrating fusion of superhuman sound and color. These swords of fire, these pools of blue-orange lava, these shooting stars: this is the tangled skein, these are the rainbows!

(Messiaen, quoted in Rischin, *For the End of Time*, 57)

33 John Cage, *Silence* (Middletown, CT: Wesleyan University Press, 1961), 19–20.
34 Olivier Messiaen, Claude Samuel, *Music and Color: Conversations with Claude Samuel*, trans. E. Thomas Glasow (Portland, OR: Amadeus Press, 1994), 34; quoted in Andrew Shenton, "Observations," 178. Messiaen reportedly believed that music created its own time in opposition to the chronometric register against which every presentiment of eternity struggles; see 182.
35 Samuel, *Music and Color*, 34; quoted in Shenton, "Observations," 178. See Ian Darbyshire, "Messiaen and the Representation of the Theological Illusion of Time," *Messiaen's Language of Mystical Love*, ed. Siglind Bruhn (New York: Routledge, 1998), 437 ff.
36 Pople, *Messiaen*, 44. Pople points out that "Messiaen himself indicated when he conceded in his conversation with Antoine Goléa that the title of the *Quatuor* might contain a play on words concerning on the one hand the end of mortal time and on the other his hope for an end to 'the equal and measured time of classical music.'" See Darbyshire, "Messiaen and the Representation of Theological Illusion in Time." Darbyshire quotes Messiaen:

> in any change there is successiveness, one part coming after another, and by numbering the antecedent and consequent parts of change there arises the notion of time, which is simply the numberedness of before and after in change.... [Consequently,] awareness of invariability in something altogether free of change produces the notion of eternity.
>
> (41–42)

From this vantage point, the beat gives birth to time by dividing time into before and after; palindromic rhythmic structures confuse their ordered succession according to the metaphor of the arrow of time. Formally speaking, the original and retrograde versions of such rhythmic structures are the same. See also Andrew Shenton, "Five Quartets," 153 ff. Andrew Shenton stresses that "Messiaen defines both time and eternity by quoting St. Thomas Aquinas, whose discussion in the *Summa Theologiæ* makes the distinction that 'Eternity is an all-encompassing simultaneity, and in time there is a before and an after'" (154). According to Shenton, Messiaen like Eliot has "a sense that the cessation of time may be a type of heaven or a unity with a god who is without beginning or end" (153).

See also Olivier Messiaen, "*Préface* to *Quatuor pour la fin du Temps*," in *Quatuor pour la fin du Temps: pour violon, clarinette en si bémol, violoncelle et piano*" (Paris: Durand Éditions Musicales, 1942), ii–iii; Olivier Messiaen, *Technique of My Musical Language*, 2. vols., trans. John Satterfield (Paris: Alphonse Leduc, 1956); Rischin, *For the End of Time*, 53; Paul Griffiths, *Olivier Messiaen and the Music of Time* (Ithaca, NY: Cornell University Press, 1985). For Griffiths, the "presence of the eternal within the temporal" (18) permeates Messiaen's oeuvre and infuses it with its truth. Like many theorists, critics, and analysts, Griffiths emphasizes the significance of Messiaen's Catholic faith. According to Griffiths, the techniques employed by Messiaen obscure or obliterate forward-directed motion (which he points out is endemic to Western music) as a way of abolishing time. Griffiths accordingly argues that the depiction of a sense of stasis is critical to the theological message of Messiaen's work.

37 Messiaen, *Quatuor pour la fin du Temps*, ii: "*Les quatre instruments à l'unisson affectent des allures de gongs et trompettes (les six premières trompettes de lApocalypse suives de catastrophes diverses, la trompette de septième angle announçe*

consommation du mystère de Dieu." Cf. Sander van Mass, *The Reinvention of Religious Music: Olivier Messiaen's Breakthrough toward the beyond* (New York: Fordham University Press, 2009). According to Van Mass,

> the positive relation between religion (that is, the three Abrahamic traditions, and mainly Judaism and Christianity) and technology, the description of technology as a form of religion, and even the description of religion as a technological enterprise, are all rooted in the Bible.
>
> (140)

Hence, for him, "it is conceivable that religion should be sought precisely in the calculative domain of technology, or more generally in the domain of repetition in various guises (animus, the concept of the artwork, notational techniques, and the like)" (141).

38 Messiaen, "*Préface* to *Quatuor pour la fin du Temps*," i; see Rischin, *For the End of Time*, 59.
39 See Paul Ricoeur, *Freedom and Nature: The Voluntary and the Involuntary*, trans. Erazim V. Kohák (Evanston, IL: Northwestern University Press, 1966), 444 ff.
40 Shenton, "Observations," 178.
41 Ernst Bloch, *Principle of Hope*, trans. Neville Plaice et al. (Oxford: Basil Blackwell, 1986), vol. 2, 818: "comparable to this, in Giotto every sacred event itself posits the place where it is located, and the division of the surface of the picture is strictly oriented by the weightiness of the event occurring within it." See Henrik Wels, "Late Medieval Debates on the Location of Angels after the Condemnation of 1277," in *Angels in Medieval Philosophical Inquiry: Their Function and Significance*, ed. Isabel Iribarren and Martin Lenz (Burlington, VT: Ashgate, 2008).
42 See Rischin, *For the End of Time*. Rischin draws on Robert Sherlaw Johnson's commentary in identifying the birdsong in *Abîme des oiseaux* as that of the blackbird. Messiaen later stated: "The bird-songs are written in the whimsical and cheerful style of the blackbird" (Messiaen, quoted in Rischin, 59).
43 Jean-François Lyotard, *The Postmodern Condition: A Report on Knowledge*, trans. Geoff Bennington and Brian Massumi (Minneapolis, MN: University of Minnesota Press, 1984), 81.
44 Lyotard, *Postmodern Condition*, 81.
45 Ibid.
46 Philip Glass, *Music by Philip Glass*, ed. Robert T. Jones (New York: Harper and Row, 1987), 29.
47 Glass, *Music by Philip Glass*, 33. Wilson envisioned the use of the opera's visual themes in

> three distinct ways: (1) a landscape seen at a distance (the Field/Spaceship scenes); (2) still lifes seen at a middle distance (the Trial scenes); and (3) portraits seen as in a closeup (the Knee Plays). As these three perspectives rotated through the four acts of the work, they created a sequence of images in an ordered scale.
>
> (33)

48 Paul Ricoeur, *Time and Narrative*, vol. 1, trans. Kathleen McLaughlin and David Pellauer (Chicago, IL: University of Chicago Press, 1984), 68. The recapitulation of a story's initial conditions by way of its "sense of an ending" governs the mode of repetition (*Wiederholung*) in which we grasp the necessary character of the incidents, episodes, and events leading to the story's conclusion.
49 Ricoeur, *Time and Narrative*, vol. 1, 67.
50 Glass, *Music by Philip Glass*, 33.

51 Paul Ricoeur, *Time and Narrative*, vol. 2, trans. Kathleen McLaughlin and David Pellauer (Chicago, IL: University of Chicago Press, 1985), 23.

52 Ricoeur, *Time and Narrative*, vol. 2, 28.

53 Glass, *Music by Philip Glass*, 33.

54 Glass, *Music by Philip Glass*, 60. This formula "occurs throughout the work, forming the basis of the Spaceship music in Act IV as well as appearing in Knee Plays 3, 4 and 5, making it one of the major thematic devices of *Einstein*" (60). See Theodor W. Adorno, *Aesthetic Theory,* ed. Gretel Adorno and Rolf Tiedemann, trans. and ed. Robert Hullot-Kentor (Minneapolis, MN: University of Minnesota Press, 1997). Adorno remarks that Stockhausen's *Zeitmaße* "evokes a through-composed cadence, a fully presented yet static dominant" (159).

55 Philip Glass, quoted in Wim Mertens, *American Minimal Music: La Monte Young, Terry Riley, Steve Reich, Philip Glass*, trans. J. Hautekiet (London: Kahn and Averill, 1983), 79. The harmonic structure of this "standing" cadential figure, which according to Glass lasts for more than eight minutes, is combined with rhythmic structures derived from additive processes in an "overall unified process" (Glass, *Music by Philip Glass*, 61–62). Each scene in *Einstein on the Beach* similarly "presents a different way of 'solving' the problem" (59) of combining harmonic sequences with rhythmic structures.

56 Jean-François Lyotard, *The Inhuman: Reflections on Time*, trans. Geoffrey Bennington and Rachel Bowlby (Cambridge: Polity Press, 1991), 163.

> What appears to the mind (the ear) of humans successively is received in one go by the divine ear. The distinction between the horizontal and the vertical in musical writing, impossible for the naked monad for lack of a support that can be inscribed, is irrelevant for the divine monad for lack of spacing. All the beatings of what we spread out in what we would call the sound-history of the world are received as in a single chord, which has neither beginning nor end.... Perhaps this celestial music should be imagined on the model of... that originary element of matter, held to itself by a terrible attraction or interaction, that astrophysics today imagines "before" its explosion into a world and into diachronic time.
>
> (162)

57 Immanuel Kant, *Critique of Judgment,* trans. Werner S. Pluhar (Indianapolis, IN: Hackett Publishing Co., 1987). Kant emphasizes that the law of reason "knows no other determinate measure that is valid for everyone and unchanging than the absolute whole" (114).

58 Jean-François Lyotard, *Lessons on the Analytic of the Sublime: (Kant's Critique of Judgement §§23–29)*, trans. Elizabeth Rottenberg (Stanford, CA: Stanford University Press, 1994), 239; see Jean-François Lyotard, *The Differend: Phrases in Dispute*, trans. Georges van den Abbeele (Minneapolis, MN: University of Minnesota Press, 1988). A "differend [*différend*] would be a case of conflict, between (at least) two parties, that cannot be equitably resolved for lack of a rule applicable to both arguments" (xi).

59 Lyotard, *Lessons*, 239. The differend "can only be felt fully in thought if the finite thought (that of form) removes itself from its finality" (151). Hence for Lyotard,

> [o]nly the Idea of absolute causality can legitimate the *double bind* in which the imagination is caught and kept prisoner in sublime feeling: to present the unpresentable. For only the Idea of absolute causality contains within it, as its very content, the obligation of its realization.... Lacking the power to present it—there being no possible presentation—the imagination will present that there is an unpresentable object, absolute causality. Thus the distress of the imagination becomes the sign of the intelligible in the sensible.
>
> (141; original emphasis)

60 Lyotard, *Lessons*, 151; see Lyotard, *The Inhuman*, 98. Cf. Kant, *Critique of Judgment*, 115.
61 Lyotard, *The Inhuman*, 155.
62 See Kant, *Critique of Judgment*. Hence the feeling of pleasure in the sublime arises only indirectly through the presentation of a form's unboundedness, to which, Kant explains, "we add … the thought of its totality" (98). For reflective judgment, the pleasure felt by reason in the presence of its own idea of the supersensible elevates the imagination beyond the abyss in which, according to Kant, it "is afraid to lose itself" (115).
63 Lyotard, *Lessons*, 127.
64 Ricoeur, *Figuring the Sacred*, 208. See Georg Wilhelm Friedrich Hegel, *Hegel's Phenomenology of Spirit*, trans. A. V. Miller (Oxford: Oxford University Press, 1977). The conditions that make freedom manifest in the world as real and rational through spirit's self-actualization are consequently a function of the philosopher's conviction that the "whole is nothing other than the essence consummating itself through its development" (11).
65 Ricoeur, *Figuring the Sacred*, 114. I want to thank Morny Joy for her remarks on this passage (pers. comm., May 15, 2016).
66 Lyotard, *The Inhuman*, 156.

> If there is no subject to refer to itself, i.e. to its power of synthesis, the sensory form and conceptual operators, so as to refer to *this* nuance, the reason is that sonorous matter which *is* this nuance is there only to the extent that, then and there, the subject is not there.... [T]his is how Epicurus circumscribes death: if it's there, I'm not there; so long as I'm there, it's not there.

Hence, if there is one thing "which does not tolerate repetition, it is death" (157).
67 Lyotard, *The Inhuman*, 157.
68 Lyotard, *The Inhuman*, 156–157: "The nuance, as non-formalized matter, escapes the syntheses, both of apprehension and of reproduction, which usually see to the grasping of sensory matter to ends of pleasure (through forms), or of knowledge (through schemata and concepts)." Hence Lyotard's negative definition of matter: "what breaks the mind" (156).
69 Lyotard, *Lessons*, 23; see Lyotard, *Postmodern Condition*. According to Lyotard, repetitive forms of contemporary music attempt to approximate the "vibratory, musical property of narrative" (21) that he identifies with the ritual performances of stories handed down in a language the meaning of which is obscure. For him, the kind of knowledge that these forms of music try to recuperate has a surprising feature:

> as meter takes precedence over accent in the production of sound (spoken or not), time ceases to be a support for memory to become an immemorial beating that, in the absence of a noticeable separation between periods, prevents their being numbered and consigns them to oblivion.
>
> (22)

Lyotard can therefore say that all genuine music aspires to "exemption from synthesis, forms, becomings, intentions and retentions, from [metaphysical] repetition, in a word" (Lyotard, *The Inhuman*, 163).
70 See my "Feeling, Interiority and the Musical Body," in *Ricoeur and the Body*, ed. Gonçalo Marcelo and Cyndie Sautereau (Lanham, MD: Lexington Books, forthcoming).
71 Eliade, *The Sacred and the Profane*, 109.
72 Ricoeur, *Time and Narrative*, vol. 2, 24.
73 Ibid.

74 Ibid.

75 Lyotard, *The Inhuman*, 156.

76 Glass, quoted in Mertens, *Minimal Music*, 88; see Steve Reich, *Writings about Music* (Halifax, NS: Press of the Nova Scotia College of Art and Design, 1974). In *Hermeneutics and Music Criticism*, I comment on how repetitive processes in which recurring motifs are slowly displaced similarly evoke the semblance of a pure, sound event. The play of differences set in motion by these processes engender the "psycho-acoustical by-products" (Reich, *Writings*, 10) emanating from this play.

77 Jacques Attali, *Noise: The Political Economy of Music*, trans. Brian Massumi (Minneapolis, MN: The University of Minnesota Press, 1985), 125. Repetitive music is accordingly the "blasphemous herald of the death of a society in which reality is only a normalized, liquidating artifice" (125–126).

78 Attali, *Noise*, 126; original in italics. For Attali, death is therefore more generally present in a repetitive economy governed by the principle of infinite exchange. The efficiency with which this economy produces the demands that it meets transforms the basic structure of the system of commodity exchange. Consequently, use-time "becomes a substitute, not a preliminary condition, for use" (101). Attali accordingly stresses that the "essential aspect of the new political economy that this kind of consumption announces is the *production of demand*, not the *production of supply*" (103; original emphasis). In the case of the musical recording, the value "of the object is not in the work itself, but in the larger whole within which the demand for commodities is constructed" (103). As "in music, repetition requires an attempt to maintain diversity, to produce a meaning for demands" (128–129) from which the pleasure of any real use-value has been expunged. See Theodor W. Adorno, *Introduction to the Sociology of Music*, trans. E. B. Ashton (New York: Continuum Publishing Company, 1976), 47; Fredric Jameson, *Postmodernism, or, the Cultural Logic of Late Capitalism* (Durham, NC: Duke University Press, 1991); Robert W. Fink, *Repeating Ourselves: American Minimal Music as Cultural Practice* (Berkeley, CA: University of California Press, 2005). Fink identifies minimalist music's significance as a cultural practice with a mode of subjectivity governed by desires engineered to create the demand met by the mass production of consumer goods.

79 Ricoeur poignantly reminds us that wisdom "fulfills one of religion's fundamental functions, which is to bind together *ethos* and *cosmos*" (Paul Ricoeur, "Toward a Hermeneutic of the Idea of Revelation," *The Harvard Theological Review* 70, no. 1/2 [1977], 12). Religion joins the sphere of human action and the world's order at the point at which they are most discordant—"in suffering and more importantly, unjust suffering" (12). Hence, wisdom

> does not teach us how to avoid suffering, or how magically to deny it, or how to dissimulate it under an illusion. It teaches us how to endure, how to suffer suffering. It places suffering in a meaningful context by producing the active qualify of suffering.
>
> (12)

80 Paul John Frandsen, "Philip Glass's *Akhnaten* (1993)," in *Writings on Philip Glass: Essays, Interviews, Criticism*, ed. Richard Kostelanetz (Berkeley, CA: University of California Press, 1997), 234; see Glass, *Music by Philip Glass*. Glass identifies Akhnaten as the "man of Religion" (139).

81 Glass, *Music by Philip Glass*, 192; see 154.

82 Glass, *Music by Philip Glass*, 173–174. The funerary images that enclose the action of this epic spectacle, which "we, like our Egyptians on stage, never quite

have out of our minds" (154), herald the endless expanse of death's timeless order. See Felix Ó Murchadha, "Being as Ruination: Heidegger, Simmel, and the Phenomenology of Ruins," *Philosophy Today* 46, SPEP Supplement (2002). Ó Murchadha underscores how ruins manifest "human powerlessness in the face of the power of time" (10). The phenomenon of decay thus reveals itself to be the essence of ruins in accordance with time's wearing effects. According to Ó Murchadha, "the ruin spooks because it is present without belonging to the present. Ruins are [therefore] neither present nor absent ...: [instead, they are] a place where everyone is a stranger" (15). I want to thank Wayne Turner for bringing this article to my attention.

83 Glass, *Music by Philip Glass*, 174; see Frandsen, "Philip Glass's *Akhnaten*."

84 Glass, *Music by Philip Glass*. This "ending" motif also appears

> before Akhnaten's Coronation in Act I, preceding the Hymn, and at the conclusion of the Hymn.... In the final moments of the opera, it finally assumes the root position with the A below the E, achieved through a final appearance of the Trilogy theme .... The Trilogy theme becomes the key that turns the lock, setting the ending firmly in A minor, the "relative minor" of C Major, the key in which *Einstein* began.
>
> (174–175)

85 Glass, *Music by Philip Glass*, 153: "Probably no culture was as death-conscious as the Egyptians."

86 Bloch, *Principle of Hope*, vol. 3, 1123.

87 Bloch, *Principle of Hope*, vol. 3, 1123, see 1122.

88 Friedrich Nietzsche, *Thus Spoke Zarathustra*, trans. Reginald John Hollingdale (London: Penguin Books, 1969), 247; quoted in Hans-Georg Gadamer, *A Century of Philosophy: A Conversation with Riccardo Dottori*, trans. Rod Coltman with Sigrid Koepke (New York: Continuum, 2003), 136.

89 Gadamer, *A Century of Philosophy*, 135.

90 Ibid., 135–136.

91 Jacques Derrida, "The Voice That Keeps Silence," in *The Sound Studies Reader*, ed. Jonathan Sterne (New York: Routledge, 2012), 501.

7 *Mimesis*, aesthetic experience, and politics

The notion that the experiences afforded by music open the world to us anew sets in relief music's performative force. Investigations and critiques in sound studies, cultural studies, and musicology that identify music's real import with social or political meanings underscore how music's expressions impact our manner and style of engaging with the world. By assigning music a social or political value, critiques that denounce the nineteenth century's metaphysical investment in music's aesthetic autonomy thus reserve a place for the claims originating from individual works. Hence, the question to which I will turn arises: To what truth do works lay claim if the experiences occasioned by them are merely aesthetic?

Over the course of my reflections and interpretations in the preceding chapters, I have argued that music revalues time in response to the deficiency born from the difference between the being and the nonbeing of time. Music, I have indicated moreover, refigures time's ultimate unrepresentability through staking out the borderlines between time and its other. In the first two chapters, I explained why music's temporal features and affective attributes, which are both attributable to the configurating operation from which they proceed, call for a philosophical consideration of the enigmas arising from the confrontation between two competing concepts of time, one cosmological, the other phenomenological. Overlooking or ignoring these enigmas occludes the productive, creative potential that I have identified with music's mimetic efficacy. My remarks on the feeling of height in Bach's music draws out one of the ways in which music replies to the enigma that springs from this confrontation between cosmological and phenomenological time. The presentiment of abiding grace that I said evinces a sense of eternity that is more Hebraic than Hellenic is accordingly one mimetic rejoinder to time's ultimate inscrutability.

By turning to the question of the truth of the claims stemming from music's expression of affective tonalities that are unique to individual works, I will in part be drawing together remarks that I made in connection with my reflections and observations on music's temporal constitution. When in Chapter 4 I claimed that Schoenberg's *Erwartung* subverts established social conventions and norms by negating tonality's ontotheological schema,

I also indicated that a work's worlding power is the condition of modern art's and music's disruptive force. Furthermore, I mentioned in passing that the "rule" exemplified by the work is the spring of its claim to truth. One of my objectives in this chapter is to show that this "rule," which the work itself summons, is a function of the operation that transforms the rectilinear succession of sounding elements into temporally meaningful complexes. In turn, I will explain how the work's exemplification of this "rule" bears out the imagination's productive role in schematizing this synthetic operation (as I anticipated in a remark in Chapter 1).

The lateral transposition of aesthetic experience onto the planes of ethics and politics brings to the fore the imagination's role in judgments that answer the demands of different situations. I postponed addressing the question of music's political value in order to give fuller consideration to the difference between a work's proposal of meaning and an injunction issuing from an exemplary moral or political act. At various points in my earlier reflections, I insisted on distinguishing between the claims made by works and religious and moral injunctions. Moreover, in Chapter 5, I indicated that the task of this final chapter in part will be to consider how music's mimetic refiguration of time's ultimate unrepresentability affects our understanding of the force of the present. The philosophical and political challenges posed by the two great aporias that Ricoeur identifies in his conclusion to his three-volume *Time and Narrative* converge on the question of history's temporalization. When in Chapter 6 I cautioned against collapsing the distinction between moral and political injunctions and a work's proposal of a manner of inhering in the world, I reserved a place for addressing these two aporias within the context of my overarching thesis. To be sure, a fuller consideration of the ethical and political implications of the aporia that arises from the fact that there is no narrative capable of meeting the requirement of a single humanity with a singular history exceeds the scope of my claim that music revalues time in answer to the aporia of time's ultimate inscrutability. Yet, to the degree that music revalues time in response to the deficiency born from the difference between time and its other, we can catch sight of this aporia and of the idea of justice's federating force from the vantage point of the series of reflections that I have undertaken, the rhythm of which traces out the mimetic exchanges between music's temporally affective value and the fragility of our mortally finite condition.

An enigmatic distance

The idea that a work's distance from the practical order of everyday life is a function of music's and art's social emancipation poses a formidable obstacle when it comes to accounting for a work's capacity to subvert ideologically congealed conventions and habits of thought. The romantic belief in music's metaphysical dignity, which rests on the nineteenth-century conviction that the aesthetic constitutes a realm of spiritual freedom isolated from the exigencies of material life, has been widely rejected. Correlatively, the metaphysical value of music's transcendent power, which served as the mark of bourgeois society's cultural superiority, has been roundly denounced. As the mark of a life liberated from material wants and needs, the spiritual formation (*Bildung*) afforded by an aesthetic education has come to be regarded as a function of music, literature, and art's institution as aesthetically autonomous. Insulated from the world of everyday concerns and affairs, cultural life from a critical sociological standpoint is accordingly the privileged sign of a social disposition that exploited music, literature, and art's conversion into an instrument of power.

The contention that any retreat from the practical order of everyday experiences masks music's, literature's, or art's ideological complicity with hegemonic systems of thought, however, is immediately caught up in an intractable paradox. On the one hand, attributing the distance separating music from the practical order of life to the aesthetics' institutionalization as a separate sphere critically targets the collusion of music's social emancipation with music's deployment as a form of cultural capital in struggles to attain advantageous social positions. On the other hand, we might wonder whether deconstructions of music's one-time sacrosanct autonomy are symptomatic of the crucial sociological event that Hannah Arendt maintains coincides with labor's elevation as the preeminent principle of human productivity.[1] The reversal of the principle of music's aesthetic autonomy into its opposite—namely, that of its ideological complicity—not only parallels the reversal that Arendt argues distinguishes the modern current of our tradition of political thought from its historical antecedents. The predominant role that the categories of production and consumption have come to play in discourses of music's roles and functions also ratifies the value accorded labor, to which Arendt argues Marx attributed a "productivity it [labor] never possesses."[2] The destruction of romantic and formalist precepts in favor of contextualizing interpretations that locate music's real meaning and significance in the material conditions of its creation, circulation, and reception are paradoxically rooted in the political soil of labor's emancipation from its traditional place within the schema of the *vita activa*.[3] Hence, the intractability of the paradox. For how can a work subvert the material conditions in which it is ideologically entangled if at the same time it only reflects or reproduces those conditions in another form?

If, as I have suggested, the strategic dismantling of music's metaphysical conceits falls prey to a reversal like the one that Arendt insists raises the category of labor to its unprecedented rank by turning the traditional hierarchy of the *vita contemplativa* and the *vita activa* on its head, the difficulty of reclaiming the aesthetics' emancipatory potential becomes all the more daunting. In their own ways, Terry Eagleton, George Levine, Peter Bürger, and Theodor Adorno have made cases for recovering music, literature, and art's critical or productive force. Eagleton, for example, emphasizes that the "imaginative reconstruction of our current practices"[4] is indispensable to avoiding the amalgam of disillusionment and sterile utopianisms that afflicts the Frankfurt school critical theorists. For him, the mystifying "escape from or sublimation of unpalatable necessity"[5] that insinuates itself behind the cultural separation of processes of fantasy and pleasure from the fulfillment of material wants is therefore only one of the functions of the aesthetic. George Levine similarly makes a plea for a view of the aesthetic that would acknowledge literature's creative possibilities.[6] Both Eagleton and Levine are wrestling with the difficulty of wresting a productive value from the aesthetics' relegation to an ideological sphere. Yet the paradox remains: For how could music, literature, or art subvert the real if the distance separating them from the material order of everyday life is at the same time the condition of their pathologically deleterious role?

Crediting music's critical social value to its aesthetic emancipation brings the aporia in question sharply into focus. In the Introduction, I indicated that replacing the metaphysics of music's transcendence with contextualizing interpretations that identify music's meaning and significance with socially constructed representations of gender, identity, race, and ethnicity inverts the schema that these interpretations ostensibly intend to combat. So long as this schema frames the riposte to absolute music's metaphysical elevation, it holds deconstructive strategies hostage to the terms of the opposition that sets music's aesthetic autonomy against its material social and political value. Within this schema, music's distance from the practical order of everyday life is an effect of the history inaugurated by Kant's transcendental justification of judgments of taste, as I explained. This distance, I will point out shortly, is for Bürger and Adorno the condition of art, literature, and music's critical, subversive force. So I ask again: How could the historical situation that is responsible for the aesthetics' ideological relegation to a separate sphere also be the condition for music, literature, and art's emancipatory potential?

According to Bürger, modernist avant-garde movements intended to combat art's isolation by capitalizing on the aesthetics' institution as a separate sphere, which the increasing division of labor and social specialization within bourgeois society had brought about. From this standpoint, art's capacity to criticize social praxis and its domination by economic interests and material forms of exploitation is a function of art's institution as a separate subsystem. Aestheticism's principle of *l'art pour l'art* thus fulfills a positive role by severing art's connection with the instrumental logic driving bourgeois society's development. On this view, the institutional framework erected by the division and specialization of different social functions releases art from the demands of this logic by removing it from the immediate context of practical exigencies. Bürger emphasizes that the "(relative) freedom of art vis-à-vis the praxis of life is at the same time the condition that must be fulfilled if there is to be a critical cognition of reality."[7] Hence, art's express lack of a social function, which aestheticism purportedly safeguards in accordance with art's claim to its own autonomous standpoint, ostensibly defines art's intended purpose by setting this purpose over against the instrumental calculus of social life.

Art's detachment from social praxis thus seems to found the space in which the imagination is free to explore the possibilities that it discovers or invents. Bürger stresses that an "art no longer distinct from the praxis of life but wholly absorbed in it will lose the capacity to criticize it, along with its distance."[8] Art's distance from the practical order is therefore "requisite for that free space within which alternatives to what exists become conceivable."[9] The attempt by the avant-garde to reintegrate art into the processes of life consequently proved to be a profoundly contradictory enterprise. By opposing bourgeois society's instrumental systems of organization, the historical avant-garde intended to make whatever "conflicts most strongly with the means-ends rationality … life's organizing principle."[10] Yet in attempting to reverse the processes through which experiences in society were increasingly hollowed out by the specialization of separate social subsystems, the historical avant-garde had to become part of the institution of art in order to preserve its critical function.

The failure of the historical avant-garde to sublate the difference between art and the praxis of life (the separation of which is the decisive feature of art's aesthetic autonomy in bourgeois society) is indicative of the conflicted situation diagnosed by Bürger. By projecting the image of a more perfect order, art protests against the prevailing one. Conversely, imaginary projections of a better world can also relieve society of the pressures and forces that call for change. Furthermore, Bürger's assessment of art's twofold character identifies art's distance from the forces and circumstances of society's production and reproduction with an element of freedom that is distinctly different from the freedom that inheres in the power to act. For where the latter manifests itself in the initiatives taken by social and historical agents to bring about changes through intervening in the world's course, art's explorations of imaginative alternatives, which take place in aesthetic realms of the *as if*, contain an "element of the noncommittal and an absence of any consequences."[11]

Identifying music's and art's fictive projections with their explorations of alternatives that have no existing referents in reality brings to the fore the utopian impulse animating efforts to reclaim the aesthetics' productive potential. The power of invention at work in literary works, works of art, and musical compositions is clearly an engine for reforming or revolutionizing praxis. We cannot overlook the possibility that music and art fuel unfulfilled demands by subverting congealed practices and habits of thought. James R. Currie is rightly suspicious of the idea that an aesthetic medium can harbor a political content that constitutes or produces a political act. At the same time, he contends that music's intensive politicization only masks our impotence with respect to current institutional structures.[12] But then what should we make of the critical value of music's and art's disruptive power? Music's and art's critical vehemence springs from the capacity they have to augment the practical field of our experiences through refashioning it from within. Music's and art's imaginative projections are accordingly the source of their ontological force. The notion that art's liberatory potential originates with its differentiation from the praxis of life in bourgeois society thus only obscures the efficacy that music and art have in searching out imaginative alternatives that impact our manner of feeling, thinking, and acting.

Identifying the realm of the imaginary with the distance instituted via art's aesthetic emancipation obfuscates the complex interweaving of ideology's and utopia's productive and pathological functions.[13] The aesthetics' alignment with politics subsequently lends itself to deploying music's affective force as a means of achieving an expressly political purpose. Barry Shank highlights the ensuing connection between the experience of beauty in music and the listeners' recognition of a utopic vision that has the weight of a moral and political imperative. On Shank's account, social and political anthems, which he points out have their roots in sacred songs, thus transform "crowds into mobs, mobs into movements, movements into parties, parties into nations"[14] by rousing them. It may be that infusing intimate publics (which Shank explains is "Lauren Berlant's term for groups formed in and through a sense of shared and ordinary feeling"[15]) with a quasi-religious sense of purpose and destiny induces a sense of equality among individuals. This fusion of a quasi-religious sense of ordained purpose with political activism in popular and national anthems consequently turns the triadic relation between religion, authority, and tradition, which Arendt identifies with the Roman experience of politics, toward its own ends.[16]

In contrast, Adorno highlights the value of music's and art's aesthetic emancipation in combatting their tendentious politicization. Accordingly, his suspicion of music's or art's affirmation of a better world seizes on the aesthetically autonomous work's distance from the existing social order (of which, he maintains these works are also a part) as the emblematic condition of the work's truth regarding the existing order's substantive deficiency. Adorno's conviction that utopia consists "essentially in the determined negation … of that which merely is"[17] consequently subjects every proposal of an alternative to the same radical critique that traps negative dialectics in its own performative contradiction. The hermeneutics of suspicion unleashed by this critique's relentless negativity thus places music's and art's power to impact reality in creatively productive ways at a seemingly infinite distance.

For Adorno, committed art by definition renounces all tendentious political allegiances for the sake of the world that it hopes to save. The polemical alienation that Brecht invented as a theorist, and which according to Adorno he practiced less and less as an artist "as he committed himself more firmly to the role of a friend of mankind,"[18] is therefore the standard bearer of committed art. Rather than "generate ameliorative measures, legislative acts or practical institutions"[19] that can be deployed as political instruments, authentic works of art critique the material conditions and ideological exigencies that perpetuate the individual's alienation in modern society by means of their formal constitutions. From Adorno's standpoint, a musical work or a work of art only resists the forces that deform social relations among human beings through its internal construction. Yet art for art's sake, which is the hallmark of aestheticism's claim regarding art's autonomy, "denies by its absolute claims that ineradicable connection with reality which is the polemical *a priori* of the very attempt to make art autonomous from the real."[20] Contrariwise, politically committed art cancels the distance that vests art with its critical function in the first place. Adorno's unremitting suspicion of a false consciousness forces music and art into the recesses of this paradox. As the "social antithesis of society,"[21] art becomes a refuge for truth by reason of its determinate negation of the empirically existing order.

The paradox that works of art hold out the hope of the world's transformation only through "pointing to a practice from which they abstain: [that is,] the creation of a just life"[22] underscores the intractable difficulty of crediting art's truth to its social emancipation. By insisting that art's distance from reality is art's first social characteristic, Adorno inscribes negative dialectics' performative contradiction at the heart of his aesthetic theory. Since a work becomes socially relevant in a critical sense only through constituting itself "in relation to what it is not,"[23] the truth of the work takes refuge in the work's determinate negation of ontic practices. Art's enigmatic promise that its truth preserves the possibility of the world's transformation is consequently also the vanishing point of the work's critical force.

Although Adorno insists that "what is true in art is something nonexistent,"[24] wresting art's utopian impulse from the aesthetics' social institution ultimately voids music's and art's mimetic power. For Adorno, the recollection in art of the possibility of an as yet unrealized condition of freedom "concretizes utopia without betraying it to existence."[25] Yet, by projecting the semblance of the true "at the price of [any] real reconciliation,"[26] art withdraws into the recesses of the aporia from which this semblance of truth cannot escape. Not even the revolt against the concept of harmony could rescue art's truth from the performative contradiction that inheres in Adorno's conception of art's semblance character. Hence, art's melancholic condition: Confined to "the imaginary reparation of the catastrophe of world history,"[27] art remonstrates against the world in order to recall a condition of freedom that has not yet, and may never, come to pass.

How, then, does the difference between a work and the practical order of everyday life make the former a work of art in the first place, as Adorno maintains? If, in saving the distance that is the condition of its truth, a work's power to refashion the real falls prey to a strategy where the suspicion of ideology has become total, the paradox—that this distance was won through art's aesthetic emancipation—collapses under the weight of this paradox's vicious circularity.[28] Lambert Zuidervaart rightly calls attention to this paradox when he points out that the principle of art's autonomy serves as the precondition of art's emancipatory potential at the same time that it renders art ideological.[29] Not even Adorno's twofold characterization of the work of art could disentangle art's subversive force from this paradox's paradigmatic stranglehold. However much Adorno insists that art's "double character as both autonomous and *fait social* is incessantly reproduced on the level of its autonomy,"[30] the presumption that the aesthetic forces of production and the productive forces of social labor are identical and have the same teleology eclipses the imagination's creatively subversive power. Locating the "mediation of music and society... in the substructure of the labor process underlying both realms"[31] consequently only preserves the classic relation between base and superstructure at the cost of identifying art's truth with its capacity to refashion the real in accordance with the worlds projected by it.

Imagination, *mimesis*, and truth

By turning to the role of imagination in exploring alternatives that set out different models for inhering in the world, I want to show how the distance that critics and theorists rightly maintain is the condition of music, art, and literature's subversive force springs from the way that works place the real into suspense. Crediting the aesthetics' emancipatory potential to German Idealism's appropriation of Kant's concepts of genius and fine art leads to an impasse, since this alleged potential originates with the aesthetics' relegation to a separate social sphere. The paradox that music's seditious or productive political and social value depends upon the same circumstances and conditions that give rise to music's metaphysical and ideological trappings is insurmountable. In the discussion that follows, I will therefore argue that the distance that is the condition of a work's power to subvert ideologically congealed understandings and habits of thought is one that the work introduces into the heart of the real by reason of the power of thought and imagination that is at work in it.

Setting the power of imagination against entrenched oppositions between claims regarding music's aesthetic autonomy and deconstructive readings rooted in cultural studies and sociological critiques cuts across this paradox. In the Introduction, I reiterated how the schema of music's metaphysical elevation frames the reversal of the idea that music's aesthetic autonomy safeguards music from its social contamination. The hermeneutical alternative that I am going to set out disentangles the imagination's productive capacity from the effects of the history through which the bond between music's social emancipation and the cult of *Bildung* were forged. When, in the course of the following series of reflections, I identify a work's subversive force with its power to refashion ways of feeling, thinking, and acting in accordance with the alternatives projected by the work, I will credit the distance introduced into the heart of our everyday outlooks and understanding to the power that the work has to place the real into suspense. This suspension or epoché of the real is the negative condition of the efficacy the work has in exploring other ways of feeling, of living, and of inscribing our lives in the web of relations with others. This negative condition is the mark of the work's retreat from the practical order of everyday life. Through this retreat, a work shatters the existing order of the real through transfiguring the real from within.

That the retreat from the real is the condition of a work's productively subversive force brings to the fore the paradox that I am setting against the paradox that stems from deriving the aesthetics' emancipatory potential from its ideologically deleterious function. Eagleton is right to suggest that imaginative reconstructions of ways of thinking, feeling, and acting constitute sites of resistance to our current practices and modes of conduct. Works augment the everyday order of our practical experiences through proposing new ways of inhering in the world. The prejudice that the world projected by a work is the emblem of the work's metaphysical elevation thus blocks the way to a hermeneutical understanding of the imagination's power.

To be sure, imagination plays a twofold role in cultural and social life. Set within the motivational framework of ideology and utopia's dialectical relations, the social and cultural imagination is constitutive as well as bewitching. Imagination is constitutive with regard to (1) the network of mediations that comprise the cultural system based on the symbolism of a culture's mythopoetic core; and (2) the exploration of imaginative alternatives that augment reality. Conversely, imagination is bewitching when (1) dreams of unfulfilled desires lose their anchorages in the practical order of everyday life, as is the case with utopian escape; and (2) dissimulating images mask the systemic deformation of relations among human beings.[32] Here, dissimulating images also evince the reproductive imagination's role in reproducing and reinforcing ideologically congealed practices and conventions. The dialectic between ideology and utopia is accordingly insurmountable. It is as though we can "cure the folly of utopia... [only by] call[ing] upon the 'healthy' function of ideology, and [conversely] the critique of ideologies [can]... only be conducted by a consciousness capable of looking at itself from the perspective of 'nowhere.'"[33]

It is clear from this brief summary that ideology's and utopia's non-pathological functions are the condition of their deleterious ones. We could similarly say that music's power to refashion the real in new and imaginative ways is both the condition of and the dialectical antipode to its ideological dissimulations of gendered constructions of identity and of the hegemonic political agendas that it ostensibly enacts. Semiotic correspondences drawn between music's structures and processes and socially constructed representations of gender, identity, and race are second-order determinants predicated on music's alleged ideological complicity with dominant conventions and mores. Conversely, music's creative potential brings into focus the power of imagination in introducing into the heart of the real the distance that critical theorists agree is the condition of music, literature, and art's subversive force.

The idea that a work iconically augments the real through placing the real into suspense is puzzling at first. Only when the artist's experience takes the form of a problem to be resolved does the power of imagination set to work. This extraordinary power of invention resides in the artist's "capacity to respond in a singular manner to the singular nature of the question."[34] The enigma of artistic creation consequently consists in the fact that the artist's experience, which provokes or ignites this response, is itself initially incommunicable. Yet, as Ricoeur points out:

> Something of… [the artist's] experience, precisely because it has been carried by a work, is going to be able to be communicated. Her naked experience as such was incommunicable; but, as soon as it can be problematized in the form of a singular question which is adequately answered in the form of a response that is singular as well, then it acquires communicability. The work iconically augments the lived experience, inexpressible, incommunicable, closed upon itself. It is this iconic augmentation, as augmentation, that is communicable.[35]

In the wake of the destruction of psychologizing and Romantic conceits, the idea that the work iconically augments the artist's lived experience opens the way to accounting for the experience occasioned by the work vis-à-vis the response to the question, problem or crisis as apprehended by the artist. We therefore need not only resist the temptation to regard a work as the conduit of the artist's expression of her innermost thoughts and feelings, but, rather than succumb to the intuitionalist fallacy, we need also to recognize that the only privilege the author enjoys when it comes to interpreting her work is that she is in the position of being its first listener, reader, or spectator.

By placing the work's communicability at the forefront of my reflections on the work's worlding power, I am also emphasizing the imagination's productive role on the side of the listener's apprehension of the work. In *Hermeneutics and Music Criticism*, I attributed the work's expression of feelings and moods to its exemplification of them. I said then that a work refigures affective dimensions of our experiences by giving voice to feelings and moods that the work possesses. By turning to the work's communicability, I want now to account for the enigma of artistic creation in the context of my broader concern with music's mimetic responses to time's deficiency with respect to the other of time.

Music's communicability, I would say now, paradoxically turns on the work's singular character. The work's adequacy vis-à-vis the fit evinced by the way that the work's constitutive components cohere in response to a question or problem is the mark of the work's success. (The adequation between the "singularity of the solution… and the singularity of the crisis situation"[36] as apprehended by the artist accordingly merits the term "style.") One of the archaic definitions of the word "fit" in the Oxford English

Dictionary (OED) is a "piece of music; [or] a strain." The OED also defines a strain as a "musical sequence of sounds; a melody, a tune." This definition not only provides etymological support for my claim regarding music's temporal expression, but it also points to the fact that the communicability of the feelings and moods evoked rests in part on the listener's ability to hear a succession of individual tones as comprising an air. When in Chapter 2 I attributed the evocation of feelings and moods to the operation that transforms successive tones into temporally expressive figures, I argued that the listener apprehends a tune through grasping the way that this figure coheres. The tune thus owes its communicability to the synthetic operation schematized by this figure's singular fit.

Ricoeur's account of the metaphorical operation is instructive in this regard. The predicative assimilation of non-literal attributes in a metaphorical statement produces the image that arises from the metaphorical statement's literal ruins. The initial predicative impertinence leads to the emergence in language of a new meaning through bringing into proximity two terms belonging to different semantic fields. Likening a peace process to a boxer who is all but defeated enables us to see the peace process as "on the ropes." This image is the concrete milieu in which the meaning schematized by the predicative assimilation can be heard, felt, and read. Accordingly, this icon is the matrix of the new predicative pertinence that resolves the initial semantic clash. The matrix of metaphorical attribution displays the intended image in the thickness of the imagining scene. The icon of this image thus constitutes the matrix that sustains the dynamic tension between a figurative meaning that resolves the predicative impertinence and the metaphor's literally nonsensical meaning.[37]

The enigma of iconic presentation springs from the way that an emergent meaning arises from the predicative impertinence that sets the imagination to work. Imagination, Ricoeur tells us, "is the apperception, the sudden insight of a new predicative pertinence, specifically a pertinence within impertinence."[38] Accordingly, we resolve the enigma of iconic presentation each time that we grasp the intended meaning depicted by the icon.[39] Just as a plot "functions as the narrative *matrix*"[40] of the stories that we tell, the icon is the "schematization of metaphorical attribution."[41] In short, the schematization of the new predicative pertinence is the engine of the metaphor's redescription of reality in the light of the heuristic fiction displayed in the thickness of the metaphoric scene.

The consequences of identifying the irruption of new meanings with the power of thought and imagination at work in these meanings' figuration are far reaching. First, to the degree that the world projected by a work shatters existing outlooks and habits of thought, the work's proposal of the meaning expressed by the work is the spring of the claim that the work makes. The force of this claim is borne out when, through distancing itself from the practical order, a work refashions reality from within. Ricoeur accordingly emphasizes that if

> one can speak of truth in relation to the work of art, it is to the extent that this designates the capacity of the work of art to break a path in the real by renewing the real *in accordance with the work itself*.[42]

Far from reproducing commonplace perspectives, evaluative standards, and affective dispositions, the expression of feelings and moods, the figuration of the density of experience, and the display of the pathos inscribed at the heart of the human condition is a creative rendering of established standards, conventions, dispositions, and the like. Drawing on Aristotle, Ricoeur stresses that *mimesis* "produces what it imitates."[43] Through inventing or discovering dimensions of experience that are unique to them, works enlarge the field of our experiences through exploring imaginative alternatives that ignite the passion for the possible through fueling as yet unfulfilled demands.

A second consequence accompanies the first. The work's power to refigure the practical field of our experiences stands as proof of the fact that what counts as real "always stands in a horizon of desired or feared or, at any rate, still undecided future possibilities."[44] It is difficult to see how social or political actors could intervene in the world's course if this course was in some way already fixed or determined in advance. Ricoeur similarly points out that the real "is everything already prefigured that is also transfigured."[45] The practical field of our experiences stands between two horizons, namely the horizon of a past that has been surpassed and the horizon of our future expectations. The power that works have by virtue of their capacity to propose different ways of thinking, feeling, and acting consequently overruns our concepts of reference, the real, and truth.[46] *Mimesis* not only explodes the classical concept of truth as adequation—the "accomplice of the representative illusion."[47] It is also doubtful whether substituting a concept of truth as manifestation (*aletheia*) for that of truth as adequation suffices with respect to "what *mimesis* demands of our thinking about truth,"[48] as Ricoeur points out. How, we could therefore ask, does the truth to which a work lays claim rise to the level of the universal that Aristotle says poetry teaches us when this truth is one that is as much invented as discovered?[49]

This question marks a critical turning point in my entire undertaking. In the next section, I will identify this universal with the "rule" summoned by a work. This "rule," I will argue, is one that the work exemplifies in response to the question, problem, or crisis for which the work provides the solution. The exemplary value of this "rule," I will argue moreover, has its analogues on the planes of ethics and politics. Far from collapsing the distinctions between them, aesthetic experience's lateral transposition onto the planes of ethics and politics preserves the differences between the force of the claims stemming from individual works and that of the injunctions issuing from exemplary moral and political acts.

By insisting that "human action is 'effectively' refigured through the fact of being configured,"[50] Ricoeur emphasizes that the poetics of narrativity continually draws on ethics. As thought experiments conducted in the "great ethical laboratory of the imaginary,"[51] literary fictions transvalue everyday actions by means of their evaluative textures. The evil that is the cause of the suffering inflicted by human beings on other human beings and the grief that is reborn through time's corroding power thus figure among the themes that make up these literary fiction's practical-pathetic worlds.

Music is unique in this regard. The idea that music differs from the art of narrative with regard to the latter's mimetic relation to the field of action (*praxis*) brings to the fore music's distinctive significance with regard to the experiences of passivity that I have related to the difference between time and its other, and to the absence of aseity in existence. These experiences, I have said on several occasions, attest to the alterity of an order that exceeds our capacity to grasp or master. The metamorphosis of everyday emotions (or first-order feelings), which Aristotle identifies with tragedy's cathartic force, attests to music's mimetic power. Just as the tragic poem "as thought (*dianoia*) … displays specific feelings"[52] that are unique to it, music's exemplification of feelings that it possesses also renews our manner of inhering in the world. A poem "is like a work of music in that its mood is exactly coextensive with the internal order of symbols articulated by its language"[53] in this regard. Just as in "the inversion of ordinary language, poetic language is not directed outwards, but inwards towards an interior, which is nothing other than the mood structured and expressed by a poem,"[54] music's retreat from the real unleashes its affective power. This power stands as a rejoinder to the enigma of time's inscrutable character. In answer to this enigma, music refigures time's ultimate unrepresentability through exploring regions of feelings to which individual works singularly give voice.

Aesthetic experience, exemplarity, and the hermeneutics of testimony

Music's mimetic responses to the difference between time and its other comprise a universe of affective tonalities that bear out music's force and power. When in the course of my reflections I highlighted the element of passivity evidenced by time's ultimately inscrutable character, I identified music's explorations of different regions of feeling with the fundamental experience of the pathos inscribed within our mortally finite condition. My itinerary consequently followed the rhythm of the series of reflections that were devoted to drawing out the constitutive relation between music's affective attributes and the difference between the being and the nonbeing of time. The presentiment of abiding grace to which I attributed the sense of eternity in Bach's music contrasts starkly with the crisis that in Schoenberg's *Erwartung* is comparable to the event brought about by the loss of confidence in a philosophy that dares to elevate thought about time to the level of the absolute. Similarly, Carolan's lament for his friend Charles MacCabe sets expressions of sorrow and grief that call for the work of mourning against eschatological and apocalyptic representations of the "end of time" in *Quatuor pour la fin du Temps*, *Einstein on the Beach*, and *Akhnaten*. Each work in its own way raises its claim to meaning in response to the deficiency born from the difference between time and the other of time. Hence, each work revalues time in answer to the difference between time and its other.

The communicability of the experiences occasioned by music is consequently the requisite condition of music's affective power. When I said that a tune owes its communicability to the operation that transforms a series of tones into a meaningfully expressive figure, I indicated how the fit of the tune rests on the listener's apprehension of a succession of tones as they cohere. (We would be mistaken if we were to identify this apprehension of the fit produced by schematizing this synthetic operation with the form of structural listening championed by Adorno; structural listening reifies the schema that it substitutes for a work's temporal matrix.[55]) To be sure, the communicability of the experience occasioned by a work draws on the listener's ability to hear successive tones and harmonies as forming meaningful patterns or figures that have distinctive features and attributes. Furthermore, the exercise of this ability is couched in the practical competencies that different listeners have with respect to various types of music, styles, genres, and cultural traditions. Nevertheless, it is difficult to imagine that music—or in Blacking's vocabulary "humanly organized sound"[56]—would have any expressive value at all if its formative temporal characteristics simply dissolved into atomistic "now" points. Consequently, like the claim stemming from the work's expression, the communicability of the experience in which the force of this claim makes itself felt takes hold in an act of imagination that is manifest in schematizing the synthetic operation that draws successively sounding elements into temporally expressive configurations.

The force of the claims stemming from music's expressions of its affective attributes puts into play the kind of judgment that Kant identifies as aesthetic. When Kant allowed for a "split within the idea of subsumption"[57] by differentiating between aesthetic judgment and determinative judgment, he reversed the act that places the individual case under a universal rule. Hence, in aesthetic judgment, "one 'seeks' the appropriate rule under which to place the singular experience."[58] This judgment is "'merely' reflective because the transcendental subject does not determine any universally valid objectivity, but instead only takes into account the procedure the mind follows in the operation of subsumption, proceeding in a way from below to above."[59] Consequently, in judgments of taste, the singular case summons its "rule" by exemplifying it.

The paradox—that the universal that poetry teaches (according to Aristotle) originates with a work's singular response to a problem, question, or crisis as apprehended by the composer, author, or artist—takes us to the heart of the enigma of aesthetic experience. Some of the most intractable difficulties for any critique of music, literature, or art springs from the fact that each listener's, reader's, or spectator's experience is unique. The singularity of this experience is evident when we consider that each encounter with a particular work involves an individual listener, reader, or spectator. Additionally, the meaning that a work has with regard to the experience afforded by the work defies its conceptual recuperation. How, then, could the truth to which a work lays claim rise to the level of the universal in the absence of conceptual constructs?

The exemplarity of the "rule" summoned by a work provides the key. The truth to which a work lays claim is attributable only to this "rule." As such, the "rule" exemplified by the work is the spring of the work's subversive power. The work's proposal of meaning thus not only contests congealed modalities of feeling and habits of thought through modulating our affective affinities with the world; the "rule" summoned by a work in answer to a question, problem, or crisis also constitutes the universal that instructs us in a way of being that only *this* work teaches. The singularity of the experience occasioned by the work is the *conditio sine qua non* of the work's affective force. Accordingly, the singularity of this experience is the impetus of the work's efficacy and the source of its claim to truth. The paradox that the work's imaginative venture is the source of the work's claim to truth places the work's communicability at the heart of this claim's performative force. For the work exemplifies the universal to which it lays claim only through schematizing the "rule" that the work summons. Attributing the work's expression of this "rule" to the configurating operation consequently accounts for the work's singular value at the price of accentuating the major difficulty of aesthetic experience. For in each case, the experience occasioned by the work is unique.

Aesthetic experience's lateral transposition onto the planes of ethics and politics brings to the fore the role of judgment in apprehending the universal to which a work lays claim. Earlier, I indicated that aesthetic experience's lateral transposition onto the planes of ethics and politics brings to the fore the imagination's role with respect to the power of judgment. Just as the artist's solution to a question or problem as she perceives it sets the listener's, reader's, or spectator's imagination to work, exemplary moral and political acts attest to the authors of these acts' power of invention in responding to the demands of a situation. The beauty that Ricoeur remarks is "specific to the acts that we admire ethically"[60] is proof, so to speak, of the act's fittingness in this respect. What is more, the suitability of the act vis-à-vis the problem or crisis confronting the agent evinces its exemplary value even in those difficult circumstances in which we are compelled to decide between the lesser of two bad alternatives. The exercise of good judgment in these difficult situations demonstrates the virtue that Aristotle called *phronesis*.[61] *Phronesis*, we could therefore say, is akin to an act of genius in that the solution constituted by the act is one that the act's author invents as much as discovers.

The transposition of aesthetic experience onto the planes of ethics and politics highlights the comparable relation between a work's or an act's exemplary value and the work's or act's power to refashion the real in response to a problem or crisis. If we are serious about drawing out connections between aesthetic experience and practical reason's ethical and political ambitus, we need to take account of the way that a work's singularity and communicability are the spring of its claim to truth.[62] Just as the experience of a work sets imagination to work, the example of the act establishes a model that we could imitate in the sense of following after it. Ricoeur stresses that this "effect of being drawn to follow"[63] (*Nachfolge*) in apprehending the moral act's suitability is comparable to the work's communicability. Hence, like the work's claim to truth, the injunction issuing from an individual act recommends itself to us by reason of its exemplary value.

Might the work of art provide a model for thinking about testimony? This question, which Ricoeur raises in the context of his remarks on aesthetic experience, opens a broad field of inquiry that exceeds the scope of my current undertaking. At the same time, this question marks a critical watershed that separates the force of the claims of individual works from the exemplary value of moral and political acts. Like works of art, models of goodness, courage, compassion and devotion to others in individual lives and acts provide examples that we could emulate: Mahatma Gandhi, Nelson Mandela, and Malala Yousafzai. Hence, like works of art, exemplary acts augment the practical field of our experiences by breaking a path into the heart of the real. The conjunction of the work of art's singularity and its exemplarity, which authorizes transposing aesthetic experience onto the planes of ethics and politics, is borne out by the "rule" summoned by the work. This "rule," to which I have argued the work attests through exemplifying it, is the source of the work's claim to truth. Comparable to the power of exemplary acts to refashion the real from within, this claim seeks its universality through augmenting the field of our everyday experiences in accordance with the world projected by the work.

By preserving the difference between the claims stemming from the work's fictive explorations and moral and political injunctions, aesthetic experience's lateral transposition sets the question of testimony in relief. If in my reflections I have been agnostic with respect to the plurality of meanings unleashed by music's mimetic refiguration of time's ultimately unrepresentable character, I have also insisted on relating music's affective power to the pathos born from the difference between time and its other. Music's explorations of diverse regions of feeling in response to this difference are tantamount to an avowal of the absence of aseity at the heart of our temporally finite condition. However, unlike moral imperatives, this avowal operates without injunction or constraint. How, then, could the work of art provide a model for thinking about testimony when the force of the work's claim springs from its fictive character? Conversely, how could the injunction issuing from an exemplary act lay claim to the absolute if the act itself is the singular response to the demands of a unique historical situation?

The notion that exemplary lives and acts are evidence of norms that are somehow lying in the background seems at first to confound my entire enterprise. For how can a single example testify to the absolute? Yet for Ricoeur, the "testimony given by exemplary lives ... attest[s] by a sort of short-circuit ... to the fundamental, without there being any need for them to pass through the interminable degrees of our laborious ascensions."[64] The term "testimony," he moreover tells us, "should be applied to words, works, actions, and to lives which attest to an intention, an inspiration, an idea at the heart of experience and history which nonetheless transcend experience and history."[65] For him, the irruption of the term testimony's religious meaning opens an absolutely new dimension within the semantic complex of testimony's profane senses. Conversely, through a kind of rebound effect, this new dimension at the same time augments the quasi-empirical meaning of the act of testifying manifest in the initiatives taken in answer to social, political, and historical exigencies.

The hermeneutics of testimony is therefore tantamount to a project of liberation under the name of the absolute to the degree that exemplary acts are also concrete signs of freedom's actualization. To be sure, just as Jean Nabert observes "the painter makes everything a testimony that awaits a witness,"[66] the testimony rendered by exemplary acts and lives gives something to be interpreted. By the same token, one cannot bring about the divestment of one's ego through one's own conscience "without the testimony of certain acts, certain lives, that, despite their radical contingency, their plain historicity, speak in the name of the absolute."[67] The originary affirmation that Ricoeur opposes to the claim to absolute knowledge thus opens the way to the reciprocal relation between the "promotion of consciousness and the recognition of the absolute in its signs."[68] Yet could we recognize the absolute as revealed in individual signs apart from the singular value of lives and acts that exemplify the convictions and ideals to which we can also lay hold?

Setting the singularity of exemplary lives and acts against the claim to absolute knowledge brings to the fore the difference between two dialectically related poles within the structure of action. When in Chapter 4 I introduced the idea of the practical task of actualizing freedom within our historical reality, I indicated how the concept of freedom's actualization shatters the hubris of a style of thinking that dares to elevate itself through totalizing time in an eternal present. Hegel's system recapitulates the accomplishments achieved by writing the history of these accomplishments from the end to the beginning. In contrast, linking reason's demand for totality to the impossibility and even the prohibition of positing any such totality in advance reserves a place for as yet unfulfilled claims. To the extent that exemplary acts and lives testify that other ways of conducting our lives and of living together are possible, they fuel as yet unfulfilled demands for equality, recognition, and justice by providing handholds for hope.[69] Consequently, the horizon of expectations illuminated by these exemplary acts and lives is as much a part of the field of action (*praxis*) as is the space between this horizon and a past that has already been surpassed.

The notion that exemplary lives and acts divest the ego of its self-certitude brings to the fore the aporia that animates the practical task of making freedom a reality. At the beginning of this chapter, I pointed out that we catch sight of this aporia in the mimetic exchanges between music's affective character and the ontological deficiency that strikes at the heart of our experiences of time. The aporia of the oneness of time on the plane of history's temporalization thus counterpoints the aporia of time's inscrutable character. For what sense can we make of the ethical and political implications of the practical task of preserving the tension between the spaces of our experiences and the horizons of our expectations when there is no historical plot equal to the epic of humanity? The event in thinking brought about by the loss of the Hegelian philosophy of history's credibility is indicative of the fact that there is no meta-narrative that meets the idea of a single humanity with one history. Yet the federating force of the idea of justice that accompanies the task of making freedom a reality meets the practical challenge that the idea of justice's federating force sets for itself in conjunction with this task only in response to the aporia unleashed by competing and conflicting aspirations and demands. The idea of justice sanctions the right to be different. Hence, the task of making freedom a reality has as its corollary the quest for an identity and a destiny to which individuals, communities, groups, and nations each lays claim as its own.[70]

The force of the present consequently constitutes the point of contact between the aporetic idea of the oneness of time and the enigma of time's ultimate inscrutability. If on the historical plane the idea of the oneness of time runs up against the impossibility of totalizing time in an eternal present, music's refiguration of the difference between time and its other breaches the constraints that ethics and politics impose upon themselves. Aesthetic experience's lateral transposition onto the planes of ethics and politics authorizes drawing analogies between the power of exemplary acts and lives to alter and to transform the world's historical course and the heuristic value of music's fictive explorations of regions of feeling in response to the difference between the being and the nonbeing of time. It may even be that the force of the convictions that arise from staking out the borderlines between time and eternity sometimes take on a religious, ethical, or political color. Yet the work's power to come bursting into the midst of the real differs from the test to which exemplary acts and lives put our most cherished customs and beliefs.

Far from obviating the difference between music's mimetic exemplification of affective dispositions and moral and political injunctions, transposing aesthetic experience onto these other domains preserves the critical distinction between aesthetics and politics. However much music renews the force of the present through refiguring time's ultimate unrepresentability, mimetic rejoinders to the pathos born of the brevity of our mortal lives explore the boundaries between time and eternity without constraint. Whatever meanings we draw from eschatological or mythic resonances of representations of the "end of time," from the destruction of ontotheological pretensions, from plaints that give voice to feelings of grief, or from the presentiment of an abiding grace, the sublimation of time's deficiency with respect to the alterity of an order that exceeds or engulfs it is a testament to our temporally finite condition. Music's power to affect us thus takes hold in response to the absence of aseity that places its stamp on our mortal existence.

Notes

1 Hannah Arendt, *The Promise of Politics*, ed. Jerome Kohn (New York: Schocken Books, 2005), 79.
2 Arendt, *The Promise of Politics*, 80.
3 See Hannah Arendt, *The Human Condition* (Chicago, IL: Chicago University Press, 1958). Arendt identifies three fundamental activities, each of which "corresponds to one of the basic conditions under which life on earth has been given to man" (7): labor, work, and action.
4 Terry Eagleton, *The Ideology of the Aesthetic* (Oxford: Basil Blackwell, 1990), 407.
5 Eagleton, *Ideology of the Aesthetic*, 411.
6 George Levine, "Introduction: Reclaiming the Aesthetic," in *Aesthetics and Ideology*, ed. George Levine (New Brunswick, NJ: Rutgers University Press, 1994); cf. Richard Kearney, *The Wake of Imagination: Ideas of Creativity in Western Culture* (London: Hutchison, 1988). Richard Kearney has cautioned against acquiescing to a postmodern fascination with the disappearance of man, the death of imagination, and the dissolution of every philosophy of truth:

> If postmodernism subverts the very opposition between the imaginary and the real, to the point where each dissolves into an empty imitation of the other, can we still speak of imagination at all? Does imagination itself not threaten to disappear with the disappearance of man?.
>
> (359)

7 Peter Bürger, *Theory of the Avant-Garde*, trans. Michael Shaw (Minneapolis, MN: University of Minnesota Press, 1984), 50.
8 Bürger, *Theory of the Avant-Garde*, 50. According to Bürger, Aestheticism was responsible for making the "element that defines art as an institution"—namely, art's distance from the "praxis of life"—its essential content. The avant-garde intended to dispose of art's social institution as a separate subsystem by organizing a new life praxis based in artistic activity.
9 Bürger, *Theory of the Avant-Garde*, 54; see Jürgen Habermas, "Modernity: An Unfinished Project," *The Postmodern Reader*, ed. Charles Jencks (London: Academy Editions, 1992), 160.
10 Bürger, *Theory of the Avant-Garde*, 34.
11 Ibid., 50.
12 James R. Currie, *Music and the Politics of Negation* (Bloomington, IN: Indiana University Press, 2012), 175.
13 See Paul Ricoeur, *Lectures on Ideology and Utopia*, ed. George H. Taylor (New York: Columbia University Press, 1986). The regressive analysis undertaken by Ricoeur uncovers the more complex relations between these two phenomena. By asking how a social interest can be expressed in "a thought, an image, or a concept of life" (10), Ricoeur accounts for the fundamentally symbolic structure of all social life by showing how the concept of ideology inherited from Marx is grafted onto the ideological phenomenon's integrative function. Accordingly, no

culture exists without a system of symbolic representations through which members of the group articulate their social experiences, their positions in society, and their feelings and dispositions. Drawing on Max Weber's notions of order (*Ordnung*) and domination (*Herrschaft*), Ricoeur attributes the ideological phenomenon's legitimating function to the discrepancy between a ruling authority's claim to legitimacy and the belief in a ruling authority's legitimacy. The system of legitimation consequently also constitutes a form of motivation in which the opacity of the "relationship between an interest and its expression in ideas" (95) conceals the process through which the ideas linked to particular interests appear as universally valid. Ideology's legitimating function aims to justify the claim on the part of a governing authority to its right to rule. This legitimating function is thus the mediate term between a system of symbolic mediations and ideology's dissimulating function. Placing the analysis of utopia in the same motivational framework reveals three corollary functions. First, utopia's eccentric role in exploring alternative ways of thinking, feeling, and conducting our lives is the dialectical counterpart of ideology's integrative function with respect to a system of symbolic mediations. Second, experimentations with alternative ways of sharing power that challenge the legitimacy of existing systems of rule correspond on the side of utopia to ideology's legitimating function. Finally, flights from reality, which are a form of pathological escape, are the counterpart to ideology's dissimulating role.

14 Barry Shank, *The Political Force of Musical Beauty* (Durham, NC: Duke University Press, 2014), 38; cf. Hannah Arendt, *The Origins of Totalitarianism* (New York: Harcourt Brace and Co., 1973). See also the debate between Theodor Adorno and Walter Benjamin over art's politicization, which Benjamin initiated in his "The Work of Art in the Age of Mechanical Reproduction," in *Illuminations*, trans. Harry Zohn, ed. Hannah Arendt (New York: Schocken Books, 1969), and to which Theodor W. Adorno responded in his "On the Fetish Character in Music and the Regression of Listening," in *The Culture Industry: Selected Essays on Mass Culture*, ed. Jay M. Bernstein (London: Routledge, 1991).

15 Shank, *Political Force of Musical Beauty*, 39.

16 Hannah Arendt, "What is Authority," in *Between Past and Future: Eight Exercises in Political Thought* (New York: Penguin Books, 1968). See Shank, *Political Force of Musical Beauty*. Shank admits that musical beauty "cannot reinforce a relation of equality among its listeners" (70). If as he maintains musical beauty "can only make those listeners aware of that demand while it implicates its auditors as its mutually responsible recipients" (70), its political force springs in part from the way that this demand fuels as yet unfulfilled claims in creating a community.

17 Ernst Bloch and Theodor W. Adorno, "Something's Missing: A Discussion between Ernst Bloch and Theodor W. Adorno on the Contradictions of Utopian Longing," in *The Utopian Function of Art and Literature: Selected Essays*, trans. Jack Zipes and Frank Mecklenburg (Cambridge, MA: MIT Press, 1988), 12; see Theodor W. Adorno, *Minima Moralia: Reflections from Damaged Life*, trans. Edmund F. N. Jephcott (London: Verso, 1974), 70.

18 Theodor W. Adorno, "Commitment," in *Aesthetics and Politics: The Key Texts of the Classic Debate within German Marxism* (London: Verso, 1977), 91.

19 Adorno, "Commitment,"180.

20 Ibid., 178.

21 Theodor W. Adorno, *Aesthetic Theory*, ed. Gretel Adorno and Rolf Tiedemann, trans. and ed. Robert Hullot-Kentor (Minneapolis, MN: University of Minnesota Press, 1997), 8.

22 Adorno, "Commitment," 194.

23 Adorno, *Aesthetic Theory*, 7–8; see Theodor W. Adorno, *Introduction to the Sociology of Music*, trans. E. B. Ashton (New York: Continuum, 1989). According to Adorno, the conversion of music's distance from society into the index of its purported functionless masks the ideological cooptation of its critical value.
24 Adorno, *Aesthetic Theory*, 131.
25 Adorno, *Aesthetic Theory*, 132. According to Adorno, "[w]hereas art opposes society, it is nevertheless unable to take up a position beyond it; it achieves opposition only through identification with that against which it remonstrates" (133). The recollection in art of a utopic condition of freedom is "the anamnesis of precisely what does not exist for-an-other" (74). For Adorno, the trace of memory in art's semblance of truth preserves in its secularized form the magic identification with nature from which the subject's mimetic comportment springs (54; see Theodor W. Adorno and Max Horkheimer, *Dialectic of Enlightenment*, trans. John Cumming [London: Verso, 1979]; Max Paddison, *Adorno's Aesthetics of Music* [Cambridge: Cambridge University Press, 1993], 58–59). Art accordingly testifies to the "possibility of the possible" in accordance with the object of art's longing—"the reality of what is not" (132).
26 Adorno, *Aesthetic Theory*, 52.
27 Ibid., 135.
28 See Jürgen Habermas, *The Philosophical Discourse of Modernity: Twelve Lectures*, trans. Frederick Lawrence (Cambridge, MA: MIT Press, 1987), 119; Adorno, *Aesthetic Theory*, 8; Adorno, *Introduction to the Sociology of Music*, 204.
29 Lambert Zuidervaart, *Adorno's Aesthetic Theory: The Redemption of Illusion* (Cambridge, MA: MIT Press, 1991), 32.
30 Adorno, *Aesthetic Theory*, 5.
31 Adorno, *Introduction to the Sociology of Music*, 206, see 207; see also Adorno, *Aesthetic Theory*. For Adorno, the

> aesthetic force of production is the same as that of productive labor and has the same teleology; and what may be called aesthetic relations of production—all that in which the productive force is embedded and in which it is active—are sedimentations or imprintings of social relations of production.
>
> (5)

See Theodor W. Adorno, "On the Problem of Music Analysis," trans. Max Paddison, *Music Analysis* 1, no. 2 (1982).
32 See note 13 above.
33 Paul Ricoeur, *From Text to Action: Essays in Hermeneutics, II*, trans. Kathleen Blamey and John B. Thompson (Evanston, IL: Northwestern University Press, 1991), 187.
34 Ricoeur et al., *Critique and Conviction: Conversations with François Azouvi and Marc de Launay*, trans. Kathleen Blamey (New York: Columbia University Press, 1998), 178.
35 Ricoeur et al., *Critique and Conviction*, 179.
36 Paul Ricoeur, *Time and Narrative*, vol. 3, trans. Kathleen Blamey and David Pellauer (Chicago: University of Chicago Press, 1988), 162. Ricoeur relates this notion of style to the role that the reader intuitively assigns to the implied author. Hence,

> [n]aming the work in terms of its author implies no conjecture about the psychology of invention or of discovery, therefore no assertion concerning the presumed intention of the inventor; it implies only the singularity of a solution to a problem.
>
> (162)

37 See Paul Ricoeur, *A Ricoeur Reader: Reflection and Imagination*, ed. Mario J. Valdés (Toronto: University of Toronto Press, 1991). The

> icon is to language what schema is to concept. In the same way that the schema is the matrix of the category, the icon is the matrix of the new semantic pertinence which is born of the collapse of the semantic kinds under the clash of contradiction.... [T]he icon is the schematization of metaphorical attribution.
>
> (126)

Hence, we read "through the icon of a relation, the relation" (127) that resolves the initial semantic clash. See Paul Ricoeur, *Memory, History, Forgetting*, trans. Kathleen Blamey and David Pellauer (Chicago, IL: University of Chicago Press, 2004). Ricoeur places the "whole of textual hermeneutics...under the theme of the increase in being applied to the work of art" (566).

38 Ricoeur, *A Ricoeur Reader*, 125.

39 Paul Ricoeur, "The Metaphorical Process as Cognition, Imagination and Feeling," in *On Metaphor*, ed. Sheldon Sacks (Chicago, IL: Chicago University Press, 1978), 148.

40 Ricoeur, *A Ricoeur Reader*, 105 (original emphasis).

41 Ricoeur, *A Ricoeur Reader*, 126:

> Metaphor is the figure of style which enables the preparatory stage to interrupt the conceptual formation because, in the metaphorical process, the movement towards the genre is arrested by the resistance of the difference and, in some way, intercepted by the figure of rhetoric. Imagination thus identified is undoubtedly the productive, schematizing imagination.

42 Ricoeur et al., *Critique and Conviction*, 173–174; original emphasis; see Paul Ricoeur, *The Course of Recognition*, trans. David Pellauer (Cambridge, MA: Harvard University Press, 2005). The work offers the spectator a space in which to "refigure his or her own expectations as a function of the models of configuration offered by plots engendered by the imagination on the plane of fiction" (100–101). The self-assimilation through feeling is accordingly a part of the commitment proper to the work's illocutionary force.

43 Ricoeur, *A Ricoeur Reader*, 138. *Mimesis*—imitation—here "is no longer a reduplication of reality but a creative rendering of it" (133). Hence *mimesis* "does not equate itself with something already given. Rather it produces what it imitates, if we continue to translate *mimesis* by 'imitation.'"

44 Hans-Georg Gadamer, *Truth and Method*, 2nd rev. ed., trans. Joel Weinsheimer and Donald G. Marshall (New York: Crossroad Publishing, 1989), 112.

45 Ricoeur, *A Ricoeur Reader*, 153.

46 See Ibid.

47 Ibid.

48 Ibid.

49 See Ricoeur, *A Ricoeur Reader*, 153: "once invented, it [the universal] is recognized as that which, in one way or another, had to be brought to language."

50 Ricoeur, *A Ricoeur Reader*, 152.

51 Paul Ricoeur, *Oneself as Another*, trans. Kathleen Blamey (Chicago, IL: University of Chicago Press, 1992), 164; see Paul Ricoeur, *Time and Narrative*, vol. 1, trans. Kathleen McLaughlin and David Pellauer (Chicago, IL: University of Chicago Press, 1984), 59.

52 Ricoeur, "Metaphorical Process," 155. Hence the "tragic *phobos* and the tragic *eleos* (terror and pity, as some translators say) are both the denial and the transfiguration of the literal feelings of fear and compassion" (156).

53 Paul Ricoeur, *Interpretation Theory: Discourse and the Surplus of Meaning* (Fort Worth, TX: Texas Christian University Press, 1976), 59.
54 Ricoeur, *Interpretation Theory*, 59; see Mikel Dufrenne, *The Phenomenology of Aesthetic Experience*, trans. Edward S. Casey et al. (Evanston, IL: Northwestern University Press, 1973). Dufrenne identifies a work's atmosphere with the ensemble of elements that produces it. Feeling is accordingly the "specific mode of apprehending" (178) the world emanating from a work.
55 See note 15, Chapter 4.
56 John Blacking, *How Musical Is Man?* (Seattle, WA: University of Washington Press, 1973), 32; also quoted in Chapter 1.
57 Paul Ricoeur, *The Just*, trans. David Pellauer (Chicago, IL: University of Chicago Press, 2000), 95.
58 Ricoeur, *The Just*, 95.
59 Ricoeur, *The Just*, 95. Ricoeur adds that

> we cannot pass over in silence the priority Kant himself gives to aesthetic judgment in relation to teleological judgment. This priority results from the fact that the natural order thought in terms of the idea of finality itself has an *aesthetic* dimension in virtue of its very relation to the subject and not to the object. Order affects us in that it pleases us. With this, aesthetic judgment is called for by teleological judgment as the first component of reflective judgment, hence as regards pure reflection.
>
> (95)

60 Ricoeur et al., *Critique and Conviction*, 182.
61 See Ricoeur, *The Just*, 154.
62 See Ricoeur et al., *Critique and Conviction*. Ricoeur accordingly stresses that "one would have to take into account the two main aspects of the work: its singularity and its communicability, with the particular form of universality that the latter implies" (182).
63 Ricoeur et al., *Critique and Conviction*, 182.
64 Ibid.
65 Paul Ricoeur, *Essays on Biblical Interpretation*, ed. Lewis Seymour Mudge (Philadelphia, PA: Fortress Press, 1980), 119–120. The quasi-empirical meaning of the act of testifying receives a new significance in a theology of testimony that preserves a "certain narrative kernel … in strict union with the confession of faith" (133). The situation of the trial, in which testimony is given and weighed, similarly acquires a new dimension within the context of the great trial in which the "kingdom and its justice is the stake of an immense contest between God and the Prince of the world" (141). The relation between witness and act also acquires a new significance with the irruption of the act's prophetic and kerygmatic dimensions. Here, as in the profane order, the "disciple is martyr because he is a witness, not the inverse" (132). When in a limit situation where the "test of conviction becomes the price of a life" (129), the sacrificial act itself constitutes the "testimony of conscience" in devotion to a cause.
66 Jean Nabert, quoted in Paul Ricoeur, *Figuring the Sacred: Religion, Narrative, and Imagination*, trans. David Pellauer, ed. Mark I. Wallace (Minneapolis, MN: Fortress Press, 1995), 117.
67 Ricoeur, *Figuring the Sacred*, 116. For Ricoeur, the

> moral conscience is first of all the protest against the unjustifiable. Its height lies in the demand for justification apart from which evil cannot be taken as unjustifiable. Its exteriority lies in the testimony of those acts whose ethical significance results from their position on the trajectory of

> "approaches to justification." ... Therefore, the hermeneutics of testimony can no more be separated from the problematic of the unjustifiable and of justification than this problematic can be unfolded outside of a hermeneutics of testimony.
>
> (118)

68 Ricoeur, *Essays on Biblical Interpretation*, 151. Ricoeur points to three indices that mark out the impossibility of any absolute knowledge. The first index of this impossibility "expresses the impotence of fixing the criteriology of the divine in a closed system" (152). The second "expresses the impotence of consciousness to bring all the signs [of the absolute] together in a coherent whole" (152). The third "expresses the impotence of identifying absolute reflection and absolute testimony itself raised to the rank of proof in the grand trial of meaning" (152–153). See Ricoeur, *Figuring the Sacred*. For Ricoeur, the

> kind of height that reflexive philosophy confesses indicates the superiority of an originary affirmation that is not numerically distinct from empirical consciousness. The adequation of this empirical consciousness with the spiritual act that constitutes it even defines the itinerary for the liberation of freedom.
>
> (113)

69 See my "The Wager of Imagination and the Logic of Hope," in *Paul Ricoeur in the Age of Hermeneutical Reason: Poetics, Praxis, and Critique*, ed. Roger W. H. Savage (Lanham, MD: Lexington Books, 2015).

70 See my "Crisis, Conflict, and the Struggle for Recognition," *Philosophy Today* 57, no. 1 (2013).

Bibliography

Abbate, Carolyn. *Unsung Voices: Opera and Musical Narrative in the Nineteenth Century*. Princeton, NJ: Princeton University Press, 1991.

Adorno, Theodor W. *Prisms*. Translated by Samuel and Shierry Weber. Cambridge, MA: MIT Press, 1967.

———. *Negative Dialectics*. Translated by E. B. Ashton. London: Routledge and Kegan Paul, 1973.

———. *Minima Moralia: Reflections from Damaged Life*. Translated by Edmund F. N. Jephcott. London: Verso, 1974.

———. *Introduction to the Sociology of Music*. Translated by E. B. Ashton. New York: Continuum Publishing Company, 1976.

———. "Commitment." In *Aesthetics and Politics: The Key Texts of the Classic Debate within German Marxism*, 177–195. London: Verso, 1977.

———. "On the Problem of Musical Analysis." Translated by Max Paddison. *Music Analysis* 1, no. 2 (1982): 169–187.

———. "On the Fetish Character of Music and the Regression in Listening." In *The Culture Industry: Selected Essays on Mass Culture*, edited by Jay M. Bernstein, 26–52. London: Routledge, 1991.

———. *The Culture Industry: Selected Essays on Mass Culture*. Edited by Jay M. Bernstein. London: Routledge, 1991.

———. *Aesthetic Theory*. Edited by Gretel Adorno and Rolf Tiedemann. Translated and edited by Robert Hullot-Kentor. Minneapolis, MN: University of Minnesota Press, 1997.

———. *Philosophy of New Music*. Translated and edited by Robert Hullot-Kentor. Minneapolis, MN: University of Minnesota Press, 2006.

Adorno, Theodor W., and Max Horkheimer. *Dialectic of Enlightenment*. Translated by John Cumming. London: Verso, 1979.

Anderson, Benedict. *Imagined Communities: Reflections on the Origin and Spread of Nationalism*. Rev. ed. London: Verso, 1991.

Arendt, Hannah. *The Human Condition*. Chicago, IL: Chicago University Press, 1958.

———. *Between Past and Future: Eight Exercises in Political Thought*. New York: Penguin Books, 1968.

———. *The Origins of Totalitarianism*. New York: Harcourt Brace and Co., 1973.

———. *The Promise of Politics*. Edited by Jerome Kohn. New York: Schocken Books, 2005.

Attali, Jacques. *Noise: The Political Economy of Music.* Translated by Brian Massumi. Minneapolis, MN: University of Minnesota Press, 1985.
Augustine, Saint Bishop of Hippo. *Confessions.* Translated by Richard Sydney Pine-Coffin. Middlesex: Penguin Books, 1961.
Auner, Joseph. "Schoenberg's Aesthetic Transformation and the Evolution of Form in *Die glückliche Hand.*" *Journal of the Arnold Schoenberg Institute* 12, no. 2 (1989): 103–128.
———. "'Heart and Brain in Music': The Genesis of Schoenberg's *Die glückliche Hand.*" In *Constructive Dissonance: Arnold Schoenberg and the Transformations of Twentieth-Century Culture*, edited by Juliane Brand and Christopher Hailey, 112–130. Berkeley, CA: University of California Press, 1997.
Austin, William W. "The History of the Poem and the Music." In *Prelude to "The Afternoon of a Faun": An Authoritative Score, Mallarmé's Poem, Backgrounds and Sources, Criticism and Analysis*, edited by William W. Austin, 3–19 New York: W. W. Norton, 1970.
Babich, Babette. "Mousike Techne: The Philosophical Practice of Music in Plato, Nietzsche, and Heidegger." In *Between Philosophy and Poetry: Writing, Rhythm, History*, edited by Massimo Verdicchio and Robert Burch, 171–180. London: Continuum, 2002.
Bach, J. S. "Sonata: A flaute Traverso col Basso, BWV 1034." Manuscript (2 scribes), n. d. [1750–1799]. Edited by Johann Friedrich Hering (copyist). Sing-Akademie Berlin, Staatsbibliothek zu Berlin-Preußischer Kulturbesitz, Musikabteilung.
———. *Matthäus-Passion* [*The Passion according to St. Matthew*]. Edited by Max Schneider [based on the 1929 edition of George Schumann]. London: Ernst Eulenburg Limited, 1935.
———. *Johann Sebastian Bach: Das Wohltemperierte Klavier (The Well-Tempered Clavier):48 Preludes and Fugues.* 2 vols. Edited by Hans Bischoff. Translated by Alexander Lipsky. New York: Edwin F. Kalmus, (1943) 1965.
Bartel, Dietrich. *Musica Poetica: Musical-Rhetorical Figures in German Baroque Music.* Lincoln, NE: University of Nebraska Press, 1997.
Barth, Karl. *Church Dogmatics.* Vol. IV of *The Doctrine of Reconciliation.* Edited by Geoffrey W. Bromiley and Thomas F. Torrance. Translated by G. W. Bromiley. London: T. & T. Clark International, (1956) 2004.
Becker, Judith. *Deep Listeners: Music, Emotion, and Trancing.* Bloomington, IN: Indiana University Press, 2004.
Begbie, Jeremy S. *Theology, Music and Time.* Cambridge: Cambridge University Press, 2000.
———. *Resounding Truth: Christian Wisdom in the World of Music.* Grand Rapids, MI: Baker Academic, 2007.
———. *Music, Modernity, and God: Essays in Listening.* Oxford: Oxford University Press, 2013.
Beiner, Guy. *Remembering the Year of the French: Irish Folk History and Social Memory.* Madison, WI: University Wisconsin Press, 2007.
Benjamin, Walter. "The Work of Art in the Age of Mechanical Reproduction." In *Illuminations*, edited by Hannah Arendt, 217–253. Translated by Harry Zohn. New York: Schocken Books, 1969.
Berger, Karol. *Bach's Cycle, Mozart's Arrow: An Essay on the Origins of Musical Modernity.* Berkeley, CA: University of California Press, 2007.

Bible Hub. "Revelation 10: Gill's Exposition of the Entire Bible." Accessed November 11, 2016. http://biblehub.com/commentaries/gill/revelation/10.htm.

Blacking, John. *How Musical Is Man?* Seattle, WA: University of Washington Press, 1973.

Bloch, Ernst. *The Principle of Hope*. 3 vols. Translated by Neville Plaice, Stephen Plaice, and Paul Knight. Oxford: Basil Blackwell, 1986.

Bloch, Ernst, and Theodor W. Adorno. "Something's Missing: A Discussion between Ernst Bloch and Theodor W. Adorno on the Contradictions of Utopian Longing." In *The Utopian Function of Art and Literature: Selected Essays*, translated by Jack Zipes and Frank Mecklenburg, 1–18. Cambridge, MA: MIT Press, 1988.

Bonds, Mark Evan. *Wordless Rhetoric: Musical Form and the Metaphor of Oration*. Cambridge, MA: Harvard University Press, 1991.

———. *Absolute Music: The History of an Idea*. Oxford: Oxford University Press, 2014.

Bourdieu, Pierre. *Distinction: A Social Critique of the Judgment of Taste*. Translated by Richard Nice. Cambridge, MA: Harvard University Press, 1984.

Bourke, Angela. "The Irish Traditional Lament and the Grieving Process." *Women's Study International Forum* 11, no. 4 (1988): 287–291.

Bourke, Angela, Siobhán Kilfeather, Maria Lauddy, Margaret Mac Curtain, Gerardine Meaney, Máirín Ní Dhonnchadha, Mary O'Dowd, and Clair Willis, eds. *The Field Day Anthology of Irish Writings; vol. IV: Irish Women's Writings and Traditions*. Cork: Cork University Press, 2002.

Brillenburg Wurth, Kiene. *Musically Sublime: Indeterminacy, Infinity, Irresolvability*. New York: Fordham University Press, 2009.

Buchanan, Herbert H. "A Key to Schoenberg's *Erwartung* (Op. 17)." *The Journal of the American Musicological Society* 20, no. 3 (1967): 434–449.

Bunting, Edward. *A General Collection of Ancient Irish Music, Containing a Variety of Admired Airs Never before Published and also the Compositions of Conolan and Carolan, Collected from the Harpers & c. in the Different Provinces of Ireland, and Adapted for the Piano-Forte*. Vol. 1. Dublin: W. Power and Co., 1796.

Burton, Richard D. E. *Olivier Messiaen: Texts, Contexts, and Intertext (1937–1948)*. Edited by Roger Nicholas. New York: Oxford University Press, 2016.

Buszin, Walter. E. "Luther on Music." *Musical Quarterly* 32, no. 1 (1946): 80–97.

Bürger, Peter. *Theory of the Avant-Garde*. Translated by Michael Shaw. Minneapolis, MN: University of Minnesota Press, 1984.

Butt, John. "Emotion in the German Lutheran Baroque and the Development of Subjective Time Consciousness." *Music Analysis* 29, no. 1–3 (2010): 19–36.

Byrnside, Ronald L. "Musical Impressionism: The Early History of the Term." *The Musical Quarterly* 66, no. 4 (1980): 522–537.

Cage, John. *Silence: Lectures and Writings*. Middletown, CT: Wesleyan University Press, 1961.

Calvin, John. "Articles Concerning the Organization of the Church and of Worship at Geneva 1537." In *Calvin: Theological Treatises*, edited by the Rev. John Kelman Sutherland Reid, 47–55. Philadelphia, PA: Westminster Press, (1954) 2000.

Carolan, Nicholas. "Carolan's Music." In *Amhráin Chearbhalláin/The Poems of Carolan: Reassessments*, edited by Liam P. Ó Murchú, 1–11. Dublin: Irish Texts Society, 2007.

Carpenter, Alexander. "Schoenberg's Vienna, Freud's Vienna: Re-Examining the Connections between the Monodrama *Erwartung* and the Early History of Psychoanalysis." *Musical Quarterly* 93, no. 1 (2010): 144–181.
Carpenter, Patricia, and Severine Neff. "Commentary." In *The Musical Idea and the Logic, Technique, and Art of Its Presentation*, translated and edited by Patricia Carpenter and Severine Neff, 1–86. Bloomington, IN: Indiana University Press, 2006.
Carson, Cairon. "The Suffering Ducks: How Has a Belfast Flute Sound Evolved in Such a Short Time?" *The Journal of Music* (April 1, 2009). Accessed September 20, 2016. http://journalofmusic.com/focus/suffering-ducks.
Chafe, Eric. *Tonal Allegory in the Vocal Music of J. S. Bach*. Berkeley, CA: University of California Press, 1991.
Chapin, Keith, and Andrew H. Clark, eds. *Speaking of Music: Addressing the Sonorous*. New York: Fordham University Press, 2013.
Chua, Daniel K. L. *Absolute Music and the Construction of Meaning*. Cambridge: Cambridge University Press, 1999.
———. "Drifting: The Dialectics of Adorno's *Philosophy of New Music*." In *Apparitions: New Perspectives on Adorno and Twentieth-Century Music*, edited by Berthold Hoeckner, 1–18. New York: Routledge, 2006.
Clarke, David, and Eric Clarke, eds. *Music and Consciousness: Philosophical, Psychological, and Cultural Perspectives*. Oxford: Oxford University Press, 2011.
Clarke, David. "Music, Phenomenology, Time Consciousness: Meditations after Husserl." In *Music and Consciousness: Philosophical, Psychological, and Cultural Perspectives*, edited by David Clarke and Eric Clarke, 1–28. Oxford: Oxford University Press, 2011.
Clarke, Eric. "Music Perception and Musical Consciousness." In *Music and Consciousness: Philosophical, Psychological, and Cultural Perspectives*, edited by David Clarke and Eric Clarke, 193–214. Oxford: Oxford University Press, 2011.
Cook, Nicholas. *Music, Imagination and Culture*. Oxford: Oxford University Press, 1990.
———. "Introduction: Refocusing Theory." *MTO: A Journal of the Society for Music Theory* 18, no. 1 (2012). Accessed November 10, 2016. http://mtosmt.org/issues/mto.12.18.1/mto.12.18.1.cook.html.
———. *Beyond the Score: Music as Performance*. New York: Oxford University Press, 2013.
Cox, Arnie. "Hearing, Feeling, Grasping Gesture." In *Music and Gesture*, edited by Anthony Gritten and Elaine King, 45–60. Burlington, VT: Ashgate, 2006.
———. "Embodying Music: Principles of the Mimetic Hypothesis." *MTO: A Journal of the Society for Music Theory* 17, no. 2 (2011). Accessed November 9, 2011. www.mtosmt.org/issues/mto.11.17.2/mto.11.17.2.cox.html.
Cullen, Louis M. "The Contemporary and Later Politics of *Caoineadh Airt Uí Laoire*." *Eighteenth Century Ireland/Iris an dá cultúr* 8, (1993): 7–38.
Currie, Mark. *About Time: Narrative, Fiction and the Philosophy of Time*. Edinburgh: Edinburgh University Press, 2007.
Currie, James R. *Music and the Politics of Negation*. Bloomington, IN: Indiana University Press, 2012.
Dahlhaus, Carl. *Esthetics of Music*. Translated by William Austin. New York: Cambridge University Press, 1982.
———. *Foundations of Music History*. Translated by J. Bradford Robinson. New York: Cambridge University Press, 1983.

———. *Schoenberg and the New Music*. Translated by Derrick Puffett and Alfred Clayton. Cambridge: Cambridge University Press, 1987.

———. *The Idea of Absolute Music*. Translated by Roger Lustig. Chicago, IL: University of Chicago Press, 1989.

———. "Fragments of Musical Hermeneutics." Translated by Karen Painter. *Current Musicology* 50, (1992): 5–20.

Darbyshire, Ian. "Messiaen and the Representation of the Theological Illusion of Time." In *Messiaen's Language of Mystical Love*, edited by Siglind Bruhn, 33–54. New York: Routledge, 1998.

David, Hans T. and Arthur Mendel, eds., *The Bach Reader: A life of Johann Sebastian Bach in Letters and Documents*. Rev. ed. New York: W. W. Norton, 1966.

Debussy, Claude. *Syrinx: pour flûte seule*. Paris: Société des Éditions Jean Jobert, (1927) 1954.

———. "Monsieur Croche the Dilettante Hater." In *Three Classics in the Aesthetics of Music*, translated by B. N. Langdon Davies, 3–8. New York: Dover Publications, (1928) 1962.

Derrida, Jacques. "The Voice That Keeps Silence." In *The Sound Studies Reader*, edited by Jonathan Sterne, 495–503. New York: Routledge, 2012.

Dufrenne, Mikel. *The Phenomenology of Aesthetic Experience*. Translated by Edward S. Casey, Albert A. Anderson, Willis Domingo, and Leon Jacobson. Evanston, IL: Northwestern University Press, 1973.

Dunne, Tom. "Subaltern Voices? Poetry in Irish, Popular Insurgency and the 1798 Rebellion." *Eighteenth-Century Life* 22, no. 3 (1998): 31–44.

Eagleton, Terry. *The Ideology of the Aesthetic*. Oxford: Basil Blackwell, 1990.

Eliade, Mircea. *The Myth of the Eternal Return: Or, Cosmos and History*. Translated by Willard R. Trask. Princeton, NJ: Princeton University Press, 1954.

———. *The Sacred and the Profane: The Nature of Religion*. Translated by Willard R. Trask. Orlando, FL: Harcourt, 1959.

———. *Myths, Dreams, and Mysteries: The Encounter between Contemporary Faiths and Archaic Realities*. Translated by Philip Mairet. New York: Harper and Row, 1960.

Emery, Walter. "Bach's Symbolic Language." *Music and Letters* 30, no. 4 (1949): 345–354.

Erlmann, Veit, ed. *Hearing Cultures: Essays on Sound, Listening, and Modernity*. Oxford: Berg Publishers, 2004.

Fink, Robert W. *Repeating Ourselves: American Minimal Music as Cultural Practice*. Berkely, CA: University of California Press, 2005.

Frandsen, Paul John. "Philip Glass's *Akhnaten* (1993)." In *Writings on Philip Glass: Essays, Interviews, Criticism*, edited by Richard Kostelanetz, 212–239. Berkeley, CA: University of California Press, 1997.

Friedson, Steven M. *Dancing Prophets: Musical Experience in Tumbuka Healing*. Chicago, IL: University of Chicago Press, 1996.

Gadamer, Hans-Georg. *Dialogue and Dialectic: Eight Hermeneutical Studies on Plato*. Translated by P. Christopher Smith. New Haven, CT: Yale University Press, 1980.

———. *The Relevance of the Beautiful and Other Essays*. Edited by Robert Bernasconi. Translated by Nicholas Walker. Cambridge: Cambridge University Press, 1986.

———. *Truth and Method*. 2nd rev. ed. Translated by Joel Weinsheimer and Donald G. Marshall. New York: Crossroad Publishing, 1989.

———. *Praise of Theory: Speeches and Essays*. Translated by Chris Dawson. New Haven, CT: Yale University Press, 1998.

———. *A Century of Philosophy: A Conversation with Riccardo Dottori*. Translated by Rod Coltman with Sigrid Koepke. New York: Continuum, 2003.

Glass, Philip. *Music by Philip Glass*. Edited by Robert T. Jones. New York: Harper and Row, 1987.

Glass, Philip, and Robert Wilson. *Einstein on the Beach*. London: Chester Music, 2003.

Glassie, Henry. *Irish Folk History: Tales from the North*. Philadelphia, PA: University of Pennsylvania Press, 1982.

———. *Passing the Time in Ballymenone: Culture and History of an Ulster Community*. Philadelphia, PA: University of Pennsylvania Press, 1982.

———. "Tradition." *The Journal of American Folklore* 108, no. 430 (1995): 395–412.

Goehr, Lydia. *The Imaginary Museum of Musical Works: An Essay in the Philosophy of Music*. Oxford: Oxford University Press, 1992.

Griffiths, Paul. *Olivier Messiaen and the Music of Time*. Ithaca, NY: Cornell University Press, 1985.

Gritten, Anthony, and Elaine King, eds. *Music and Gesture*. Burlington, VT: Ashgate, 2006.

Habermas, Jürgen. *The Philosophical Discourse of Modernity: Twelve Lectures*. Translated by Frederick Lawrence. Cambridge, MA: MIT Press, 1987.

———. "Modernity: An Unfinished Project." In *The Postmodern Reader*, edited by Charles Jencks, 158–171. London: Academy Editions, 1992.

Haimo, Ethan. *Schoenberg's Transformation of Musical Language*. Cambridge: Cambridge University Press, 2006.

Hanslick, Eduard. *On the Musically Beautiful: A Contribution towards the Revision of the Aesthetics of Music*. Translated and edited by Geoffrey Payzant. Indianapolis, IN: Hackett Publishing Company, 1986.

Hatten, Robert S. "A Theory of Musical Gesture and Its Application to Beethoven and Schubert." In *Music and Gesture*, edited by Anthony Gritten and Elaine King, 1–23. Burlington, VT: Ashgate, 2006.

Hegel, Georg Wilhelm Friedrich. *Hegel's Aesthetics: Lectures on Fine Arts*. 2 vols. Translated by Thomas Malcom Knox. Oxford: Oxford University Press, 1975.

———. *Lectures on the Philosophy of World History. Introduction: Reason in History*. Translated by Hugh Bar Nisbet. Cambridge: Cambridge University Press, 1975.

———. *Hegel's Phenomenology of Spirit*. Translated by A. V. Miller. Oxford: Oxford University Press, 1977.

Heidegger, Martin. *Being and Time*. Translated by John Macquarrie and Edward Robinson. New York: Harper and Row, 1962.

———. *Poetry, Language, Thought*. Translated by Albert Hofstadter. New York: Harper and Row, 1971.

———. *The Concept of Time*. Translated by William McNeill. Oxford: Blackwell, 1992.

Hill, David. "The Time of the Sign: 'O Haupt voll Blut und Wunden' in Bach's St. Matthew Passion." *The Journal of Musicology* 14, no. 4 (1996): 514–543.

Henry, Edward O. "Institutions for the Promotion of Indigenous Music: The Case for Ireland's Comhaltas Ceoltoiri Eireann." *Ethnomusicology* 33, no. 1 (1989): 67–95.

Hoeckner, Berthold. *Programming the Absolute: Nineteenth-Century German Music and the Hermeneutics of the Moment*. Princeton, NJ: Princeton University Press, 2002.

Hoffman, Ernst Theodor Amadeus. "Review of Beethoven's Fifth Symphony." In *E. T. A. Hoffman's Musical Writings: Kreisleriana, the Poet and the Composer, Music Criticism*, edited by David Charlton, 234–251. Translated by Martyn Clarke. Cambridge: Cambridge University Press, 1989.

Honneth, Axel. *The Struggle for Recognition: The Moral Grammar of Social Conflicts*. Translated by Joel Anderson. Cambridge, MA: MIT Press, 1995.

Hutchinson, John. *The Dynamics of Cultural Nationalism: The Gaelic Revival and the Creation of the Irish Nation State*. London: Allen and Unwin, 1987.

Hyde, Martha M. "Musical Form and the Development of Schoenberg's Twelve-Tone Method." *Journal of Music Theory* 29, no. 1 (1985): 85–143.

Hyer, Bryan. "Tonality." In *The New Grove Dictionary of Music and Musicians*, vol. 25 (Taiwan to Twelve Apostles), edited by Stanley Sadie, 583–594. London: Macmillan Publishers Limited, 2001.

Ihde, Don. *Listening and Voice: Phenomenologies of Sound*. 2nd ed. Albany, NY: State University of New York Press, 2007.

Ingarden, Roman. *The Work of Music and the Problem of Its Identity*. Translated by Adam Czerniawski. Edited by Jean G. Harrell. Berkeley, CA: University of California Press, 1986.

"Irish Flute: Matt Molloy Plays a Slow Air and Reels," YouTube video, 5:03, posted by "Douglas Hadden," May 13, 2015, www.youtube.com/watch?v=SiBUPFBi8Tc. Accessed September 15, 2016.

Jameson, Fredric. *Postmodernism, or, the Cultural Logic of Late Capitalism*. Durham, NC: Duke University Press, 1991.

———. "The End of Temporality." *Critical Inquiry* 29, no. 4 (2003): 695–718.

Kant, Immanuel. *Critique of Judgment*. Translated by Werner S. Pluhar. Indianapolis, IN: Hackett Publishing Co., 1987.

———. *Immanuel Kant's Critique of Pure Reason*. Translated by Norman Kemp Smith. London: MacMillan and Co., 1929.

Karnes, Kevin C. *Music, Criticism, and the Challenges of History: Shaping Modern Musical Thought in Late Nineteenth-Century Vienna*. New York: Oxford University Press, 2008.

Kars, Père Jean-Rodolphe. "The Works of Olivier Messiaen and the Catholic Liturgy." In *Olivier Messiaen: Music, Art and Literature*, edited by Christopher Dingle and Nigel Simeone, 323–334. Burlington, VT: Ashgate, 2007.

Kearney, Richard. *The Wake of Imagination: Ideas of Creativity in Western Culture*. London: Hutchison, 1988.

———. "Ricoeur's Wager of the Flesh." In *Paul Ricoeur in the Age of Hermeneutical Reason: Poetics, Praxis and Critique*, edited by Roger W. H. Savage, 179–190. Lanham, MD: Lexington Books, 2015.

Keil, Charles. "Participatory Discrepancies and the Power of Music." *Cultural Anthropology* 2, no. 3 (1987): 275–283.

———. "The Theory of Participatory Discrepancies: A Progress Report." *Ethnomusicology* 39, no. 1 (1995): 1–19.

Kerman, Joseph. "How We Got into Analysis, and How to Get Out." *Critical Inquiry* 7, no. 2 (1980): 311–331.

———. *Contemplating Music: Challenges to Musicology*. Cambridge, MA: Harvard University Press, 1985.

Koselleck. Reinhart. *Future's Past: The Semantics of Historical Time*. Translated by Keith Tribe. Cambridge, MA: MIT Press, 1985.

Kramer, Jonathan D. *The Time of Music: New Meanings, New Temporalities, New Listening Strategies*. New York: Schirmer Books, 1988.

Kramer, Lawrence. "Music Criticism and the Postmodernist Turn: In Contrary Motion with Gary Tomlinson." *Current Musicology* 53, (1993): 25–35.

———. *Classical Music and Postmodern Knowledge*. Berkeley, CA: University of California Press, 1995.

———. *Musical Meaning: Toward a Critical History*. Berkeley, CA: University of California Press, 2002.

Kretzschmar, Hermann. "Anregung zur Förderung musikalischer Hermeneutik (1902) ['Suggestions for the Furtherance of Musical Hermeneutics']." In *Musical Aesthetics: A Historical Reader*. Vol. 3 of *Aesthetics in Music: The Twentieth Century*, edited by Edward A. Lippman, 5–30. Stuyvesant, NY: Pendragon Press, 1990.

———. "Neue Anregungen zur Förderung musikalischer Hermeneutik: Satzästhetik (1905) ['New Suggestions for the Furtherance of Musical Hermeneutics: The Aesthetics of Musical Compositions']." In *Musical Aesthetics: A Historical Reader*. Vol. 3 of *Aesthetics in Music: The Twentieth Century*, edited by Edward A. Lippman, 31–46. Stuyvesant, NY: Pendragon Press, 1990.

Kübler-Ross, Elisabeth. *On Death and Dying*. London: Tavistock, 1973.

Leech-Wilkinson, Daniel. "Compositions, Scores, Performances, Meanings." *MTO: A Journal of the Society for Music Theory* 18, no. 1 (2012). Accessed November 10, 2016. www.mtosmt.org/issues/mto.12.18.1/mto.12.18.1.leech-wilkinson.php.

Lessem, Alan Philip. *Music and Text in the Works of Arnold Schoenberg: The Critical Years, 1908–1922*. Ann Arbor, MI: UMI Research Press, 1979.

Lederer, Victor. *Bach's St. Matthew Passion: A Closer Look*. New York: Continuum, 2008.

Levinas, Emmanuel. *Time and the Other*. Translated by Richard A. Cohen. Pittsburgh, PA: Duquesne University Press, 1987.

Levine, George. "Introduction: Reclaiming the Aesthetic." In *Aesthetics and Ideology*, edited by George Levine, 1–28. New Brunswick, NJ: Rutgers University Press, 1994.

Lévi-Strauss, Claude. *The Raw and the Cooked: Introduction to a Science of Mythology*. Vol. 1. Translated by John and Doreen Weightman. Chicago, IL: University of Chicago Press, 1969.

Lidov, David. "Emotive Gestures in Music and Its Contraries." In *Music and Gesture*, edited by Anthony Gritten and Elaine King, 24–44. Burlington, VT: Ashgate, 2006.

Lyotard, Jean-François. *The Postmodern Condition: A Report on Knowledge*. Translated by Geoff Bennington and Brian Massumi. Minneapolis, MN: University of Minnesota Press, 1984.

———. *The Differend: Phrases in Dispute*. Translated by Georges van den Abbeele. Minneapolis, MN: University of Minnesota Press, 1988.

———. *The Inhuman: Reflections on Time*. Translated by Geoffrey Bennington and Rachel Bowlby. Cambridge: Polity Press, 1991.

———. *Lessons on the Analytic of the Sublime: (Kant's Critique of Judgement §§23–29)*. Translated by Elizabeth Rottenberg. Stanford, CA: Stanford University Press, 1994.

Lysaght, Patricia. "'Caoineadh os Cionn Coirp': The Lament for the Dead in Ireland." *Folklore* 108, (1997): 65–82.

Mac Marcais, Aindrias. "The Deserted Land." In *The Field Day Anthology of Irish Writing*. Vol. 1. Edited by Seamus Deane, Andrew Carpenter, and Jonathan Williams, 279–280. Derry: Field Day Publications, 1991.

Mann, Thomas. *The Story of a Novel: The Genesis of Doctor Faustus*. Translated by Richard and Clara Winston. New York: Alfred A. Knopf, 1961.

———. *Doctor Faustus: The Life of the German Composer Adrian Leverkühn as Told by a Friend*. Translated by John E. Woods. New York: Vintage International, (1947) 1999.

Marcel, Gabriel. "Bergsonism and Music." In *Reflections on Art: A Source Book of Writings by Artists, Critics and Philosophers*, edited by Susanne K. Langer, 142–151. Baltimore, MD: The John Hopkins Press, 1958.

Marcus, George E., and Michael M. J. Fischer. *Anthropology as Cultural Critique: An Experimental Moment in the Human Sciences*. Chicago, IL: University of Chicago Press, 1986.

Mathewes, Charles T. "Augustinian Anthropology: Interior intimo meo." *Journal of Religious Ethics* 27, no. 2 (1999): 195–221.

"Matt Molloy-Air: The Parting of Friends/Paddy Ryan's Dream," YouTube video, 5:27, posted by "Jeremy V," January 1, 2012, www.youtube.com/watch?v=3_uSji7Azz4&list=RD3_uSji7Azz4#t=97. Accessed September 15, 2016.

McCann, May. "Music and Politics in Ireland: The Specificity of the Folk Revival in Belfast." *British Journal of Ethnomusicology* 4, (1995): 51–75.

McClary, Susan. *Feminine Endings: Music, Gender, and Sexuality*. Minneapolis, MN: University of Minnesota Press, 1991.

———. "Narrative Agendas in 'Absolute Music': Identity and Difference in Brahms's Third Symphony." In *Musicology and Difference: Gender and Sexuality in Musical Scholarship*, edited by Ruth A. Solie, 326–344. Berkeley, CA: University of California Press, 1993.

———. "Paradigm Dissonances: Music Theory, Cultural Studies, Feminist Criticism." *Perspectives of New Music* 32, no. 1 (1994): 68–85.

Maus, Fred Everett. "Music as Drama." *Music Theory Spectrum* 10, (1988): 56–73.

Mellers, Wilfrid. *Bach and the Dance of God*. New York: Oxford University Press, 1981.

Mertens, Wim. *American Minimal Music: La Monte Young, Terry Riley, Steve Reich, Philip Glass*. Translated by J. Hautekiet. London: Kahn and Averill, 1983.

Messiaen, Olivier. *Quatuor pour la fin du Temps: pour violon, clarinette en si bémol, violoncelle et piano*. Paris: Durand Éditions Musicales, 1942.

———. *Technique of My Musical Language*. 2. vols. Translated by John Satterfield. Paris: Alphonse Leduc, 1956.

———. "Address Delivered at the Conferring of the Praemium Erasmianum on 25 June 1971, in Amsterdam." In Almut Rössler, *Contributions to the Spiritual World of Olivier Messiaen: With Original Texts by the Composer*. Compiled by Almut Rüssler. Translated by Barbara Dagg and Nancy Poland, 38–47. Duisburg: Gilles and Francke Verlag, 1986.

Messian, Olivier, Claude Samuel. *Music and Color: Conversations with Claude Samuel*. Translated by E. Thomas Glasow. Portland, OR: Amadeus Press, 1994.

Mongin, Olivier. *Paul Ricoeur*. Paris: Seuil, 1994.

Morgan, Robert P. "Musical Time/Musical Space." *Critical Inquiry* 6, no. 3 (1980): 527–538.

Moyaert, Marianne. "From Religious Violence to Interreligious Hospitality." In *Paul Ricoeur in the Age of Hermeneutical Reason: Poetics, Praxis, and Critique*, edited by Roger W. H. Savage, 159–178. Lanham, MD: Lexington Books, 2015.

Nancy, Jean-Luc. *Listening*. Translated by Charlotte Mandell. New York: Fordham University Press, 2007.

Newcomb, Anthony. "Sound and Feeling." *Critical Inquiry* 10, no. 4 (1984): 614–643.

———. "Schumann and Late Eighteenth-Century Narrative Strategies." *19th Century Music* 11, no. 2 (1987): 164–174.

Nietzsche, Friedrich. *Thus Spoke Zarathustra*. Translated by Reginald John Hollingdale. London: Penguin Books, 1969.

Ní Dhomhnaill, Nuala. "A Ghostly Alhambra." In *Irish Hunger: Personal Reflections on the Legacy of the Famine*, edited by Tom Hayden, 68–78. Boulder, CO: Roberts Rinehart Publishers, 1998.

Ní Mhunghaile, Lesa. "Carolan's Verse." In *Amhráin Chearbhalláin/The Poems of Carolan: Reassessments*, edited by Liam P. Ó Murchú, 43–64. Dublin: Irish Texts Society, 2007.

Nonnus, Nonnos of Panopolis. *Nonnos Dionysiaca*. Vol. 3, bk. XLII. Translated by W. H. D. Rouse, 228–268. Cambridge, MA: Harvard University Press, 1940.

Ó Canainn, Tomás. *Traditional Music in Ireland*. Cork: Ossian, 1993.

Ó Coileáin, Seán. "Oral or Literary? Some Strands of the Argument." *Studia Hibernica* 17/18, (1977/1978): 7–35.

———. "The Irish Lament: An Oral Genre." *Studia Hibernica* 24, (1988): 97–117.

Ó Crualaoich, Gearóid. "The 'Merry Wake.'" In *Irish Popular Culture, 1650–1850*, edited by James S. Donnelly, Jr. and Kerby A. Miller, 173–200. Dublin: Irish Academic Press, 1998.

Ó Laoire, Lillis. "Towards an Aesthetic of Song Performance." *Folk Life: Journal of Ethnological Studies* 54, no. 1 (2016): 49–67.

Ó Madagáin, Breandán. "Irish Vocal Music of Lament and Syllabic Verse." In *The Celtic Consciousness*, edited by Robert O'Driscoll, 311–332. Toronto: Dolmen Press, 1981.

———. *Caointe agus Seancheolta Eile/Keening and Other Old Irish Musics*. Indreabhán: Cló Iar-Chnnacht, 2005.

Ó Máille, Tomás. *Aṁráin Ċearḃalláin: Poems of Carolan Together with Other N. Connacht and S. Ulster Lyrics*. London: Irish Texts Society, 1916.

Ó Murchadha, Felix. "Being as Ruination: Heidegger, Simmel, and the Phenomenology of Ruins." *Philosophy Today* 46, SPEP Supplement (2002): 10–18.

O'Sullivan, Donal. *Carolan: The Life Times and Music of an Irish Harper*. Cork: Ossian, (1958) 2001.

Ovid. *Metamorphoses* [689–695; 704–709]. Translated by Sir Samuel Garth, John Dryden, Joseph Addison, Laurence Eusden, Arthur Maynwaring, Samuel Croxall, Nahum Tate, and John Gay. Hertfordshire: Wordsworth Editions, (1826) 1998.

Paddison, Max. *Adorno's Aesthetics of Music*. Cambridge: Cambridge University Press, 1993.

———. "Mimesis and the Aesthetics of Musical Expression." *Music Analysis* 29, no. 1–3 (2010): 126–148.

Pelikan, Jaroslav. *Bach among the Theologians*. Eugene, OR: Wipf and Stock, 1986.

Phillips, C. Henry. "The Symbolists and Debussy." *Music & Letters* 13, no. 3 (1932): 298–311.

Pople, Anthony. *Messiaen: Quatuor pour la fin du Temps*. Cambridge: Cambridge University Press, 1998.

Priest, Eldritch. "Felt as Thought (or, Musical Abstraction and the Semblance of Affect)." In *Sound, Music, Affect: Theorizing Sonic Experience*, edited by Marie Thompson and Ian Biddle, 45–64. New York: Bloomsbury, 2013.

Racy, Ali Jihad. *Making Music in the Arab World: The Culture and Artistry of Ṭarab*. Cambridge: Cambridge University Press, 2003.

Reich, Steve. *Writings about Music*. Halifax, NS: Press of the Nova Scotia College of Art and Design, 1974.

Ricoeur, Paul. *Freedom and Nature: The Voluntary and the Involuntary*. Translated by Erazim V. Kohák. Evanston, IL: Northwestern University Press, 1966.

———. *The Symbolism of Evil*. Translated by Emerson Buchanan. Boston, MA: Beacon Press, 1967.

———. *The Conflict of Interpretations*. Edited by Don Ihde. Evanston, IL: Northwestern University Press, 1974.

———. *Interpretation Theory: Discourse and the Surplus of Meaning*. Fort Worth, TX: Texas Christian University Press, 1976.

———. *The Rule of Metaphor: Multi-Disciplinary Studies of the Creation of Meaning in Language*. Translated by Robert Czerny with Kathleen McLaughlin and John Costello, S. J. Toronto: University of Toronto Press, 1977.

———. "Toward a Hermeneutic of the Idea of Revelation." *Harvard Theological Review* 70, no. 1/2 (1977): 1–37.

———. "The Metaphorical Process as Cognition, Imagination and Feeling." In *On Metaphor*, edited by Sheldon Sacks, 141–157. Chicago, IL: Chicago University Press, 1978.

———. *Essays on Biblical Interpretation*. Edited by Lewis Seymour Mudge. Philadelphia, PA: Fortress Press, 1980.

———. *Hermeneutics and the Human Sciences: Essays on Language, Action and Interpretation*. Translated by John B. Thompson. Cambridge: Cambridge University Press, 1981.

———. *Time and Narrative*. Vol. 1. Translated by Kathleen McLaughlin and David Pellauer. Chicago, IL: University of Chicago Press, 1984.

———. *Time and Narrative*. Vol. 2. Translated by Kathleen McLaughlin and David Pellauer. Chicago, IL: University of Chicago Press, 1985.

———. *Fallible Man*. Translated by Charles A. Kelbley. New York: Fordham University Press, 1986.

———. *Lectures on Ideology and Utopia*. Edited by George H. Taylor. New York: Columbia University Press, 1986.

———. *Time and Narrative*. Vol. 3. Translated by Kathleen Blamey and David Pellauer. Chicago, IL: University of Chicago Press, 1988.

———. *A Ricoeur Reader: Reflection and Imagination*. Edited by Mario J. Valdés. Toronto: University of Toronto Press, 1991.

———. *From Text to Action: Essays in Hermeneutics, II*. Translated by Kathleen Blamey and John B. Thompson. Evanston, IL: Northwestern University Press, 1991.

———. *Oneself as Another*. Translated by Kathleen Blamey. Chicago, IL: University of Chicago Press, 1992.

———. "Wonder, Eroticism, and Enigma." In *Sexuality and the Sacred: Sources for Theological Reflection*, edited by James B. Nelson and Sandra P. Longfellow, 80–90. Louisville, KY: John Knox Press, 1994.

———. *Figuring the Sacred: Religion, Narrative, and Imagination*. Translated by David Pellauer. Edited by Mark I. Wallace. Minneapolis, MN: Fortress Press, 1995.

———. "Lamentation as Prayer." In *André LaCocque and Paul Ricoeur, Thinking Biblically: Exegetical and Hermeneutical Studies*, translated by David Pellauer, 211–232. Chicago, IL: University of Chicago Press, 1998.

———. "Religion and Symbolic Violence." Translated by James Williams. *Contagion: Journal of Violence, Mimesis, and Culture* 6, (1999): 1–11.
———. *The Just*. Translated by David Pellauer. Chicago, IL: University of Chicago Press, 2000.
———. *Memory, History, Forgetting*. Translated by Kathleen Blamey and David Pellauer. Chicago, IL: University of Chicago Press, 2004.
———. *The Course of Recognition*. Translated by David Pellauer. Cambridge, MA: Harvard University Press, 2005.
———. *Reflections on the Just*. Translated by David Pellauer. Chicago, IL: University of Chicago Press, 2007.
Ricoeur, Paul, François Azouvi, and Marc de Launay. *Critique and Conviction: Conversations with François Azouvi and Marc de Launay*. Translated by Kathleen Blamey. New York: Columbia University Press, 1998.
Rischin, Rebecca. *For the End of Time: The Story of the Messiaen Quartet*. Ithaca, NY: Cornell University Press, 2003.
Rognoni, Luigi. *The Second Vienna School: Expressionism and Dodecaphony*; (*The Rise of Expressionism in the Music of Arnold Schoenberg, Alban Berg, and Anton von Webern*). Translated by Robert W. Mann. London: John Calder, 1977.
Rosen, Charles. *Arnold Schoenberg*. New York: The Viking Press, 1975.
Rouget, Gilbert. *Music and Trance: A Theory of the Relation between Music and Possession*. Translated by Brunhilde Biebuyck. Chicago, IL: University of Chicago Press, 1985.
Savage, Roger W. H. *Structure and Sorcery: The Aesthetics of Post-War Serial Composition and Indeterminacy*. New York: Garland Publishing, 1989.
———. "Being, Transcendence and the Ontology of Music." *The World of Music* 51, no. 2 (2009): 7–22.
———. *Hermeneutics and Music Criticism*. New York: Routledge, 2010.
———. "Crisis, Conflict, and the Struggle for Recognition." *Philosophy Today* 57, no. 1 (2013): 72–83.
———. "Fragile Identities, Capable Selves." *Études Ricoeuriennes/Ricoeur Studies* 4, no. 2 (2013): 64–78.
———. "Introduction: Paul Ricoeur and the Age of Hermeneutical Reason." In *Paul Ricoeur in the Age of Hermeneutical Reason: Poetics, Praxis, and Critique*, edited by Roger W. H. Savage, 1–20. Lanham, MD: Lexington Books, 2015.
———. "Judgment, Imagination and the Search for Justice." *Études Ricoeuriennes/Ricoeur Studies* 6, no. 2 (2015): 50–67.
———. "The Wager of Imagination and the Logic of Hope." In *Paul Ricoeur in the Age of Hermeneutical Reason: Poetics, Praxis, and Critique*, edited by Roger W. H. Savage, 139–158. Lanham, MD: Lexington Books, 2015.
———. "Exemplarity and the 'Law of Superabundance.'" *Budhi: A Journal of Ideas and Culture* 20, no. 2 (2016): 19–49.
———. "Feeling, Interiority and the Musical Body." In *Ricoeur and the Body*, edited by Gonçalo Marcelo and Cyndie Sautereau. Lanham, MD: Lexington Books, forthcoming.
Scruton, Roger. *Art and Imagination: A Study in the Philosophy of Mind*. London: Methuen and Co. 1974.
———. *The Aesthetic Understanding: Essays in the Philosophy of Art and Culture*. Manchester: Carcanet Press, 1983.

———. *The Aesthetics of Music*. Oxford: Oxford University Press, 1997.

Shank, Barry. *The Political Force of Musical Beauty*. Durham, NC: Duke University Press, 2014.

Schenker, Heinrich. *Free Composition (Der freie Satz): Volume III of New Musical Theories and Fantasies*. Translated by Ernst Oster. New York: Longman, 1979.

Shenton, Andrew. "Observations on Time in Olivier Messiaen's *Traité*." In *Olivier Messiaen: Music, Art and Literature*, edited by Christopher Dingle and Nigel Simeone, 173–190. Burlington, VT: Ashgate, 2007.

———. "Five Quartets: The Search for the Still Point of the Turning World in the War Quartets of T. S. Eliot and Olivier Messiaen." In *Messiaen the Theologian*, edited by Andrew Shenton, 145–162. Burlington, VT: Ashgate, 2010.

Schoenberg, Arnold. *Style and Idea: Selected Writings of Arnold Schoenberg*. Edited by Leonard Stein. Translated by Leo Black. Berkeley, CA: University of California Press, 1975.

———. *Theory of Harmony: 100th Anniversary Edition*. Translated by Roy E. Carter. Forward by Walter Frisch. Berkeley, CA: University of California Press, 1978.

———. *Erwartung: Monodram in 1 Akt für Sopran und Kammerorchester, Op. 17*. Wien: Universal Edition, (1909) 2004. Accessed September 28, 2016. www.universaledition.com.

Schorske, Carl E. *Fin-de-Siècle Vienna: Politics and Culture*. New York: Vintage Books, 1981.

Schrimshaw, Will. "Non-Cochlear Sound: On Affect and Exteriority." In *Sound, Music, Affect: Theorizing Sonic Experience*, edited by Marie Thompson and Ian Biddle, 27–44. New York: Bloomsbury, 2013.

Schutz, Alfred. "Fragments on the Phenomenology of Music." Edited by Fred Kersten. *Music and Man* 2, no. 1–2 (1976): 5–71.

Shields, Hugh. *Narrative Singing in Ireland: Lays, Balls, Come-All-Yes and Other Songs*. Dublin: Irish Academic Press, 1993.

Simms, Bryan R. "Whose Idea Was *Erwartung*?" In *Constructive Dissonance: Arnold Schoenberg and the Transformations of Twentieth-Century Culture*, edited by Juliane Brand and Christopher Hailey, 100–130. Berkeley, CA: University of California Press, 1997.

———. *The Atonal Work of Arnold Schoenberg, 1908–1923*. New York: Oxford University Press, 2000.

Skadra, Christina A. "Alfred Schutz's Phenomenology of Music." *The Journal of Musicological Research* 3, no. 1–2 (1979): 75–132.

Solie, Ruth. "The Living Work: Organicism and Musical Analysis." *19th Century Music* 4, no. 2 (1980): 147–156.

Subotnik, Rose Rosengard. *Deconstructive Variations: Music and Reason in Western Society*. Minneapolis, MN: University of Minnesota Press, 1996.

Synge, John Millington. *The Aran Islands*. Boston, MA: John W. Luce & Co., 1911.

Szendy, Peter. *Listen: A History of Our Ears*. Translated by Charlotte Mandell. New York: Fordham University Press, 2008.

———. "The Auditory Re-Turn (The Point of Listening)." In *Thresholds of Listening: Sound, Technics, Space*, edited by Sander van Maas, 18–29. New York: Fordham University Press, 2015.

Taylor, Charles. *Hegel*. Cambridge: Cambridge University Press, 1975.

The Irishpage.com. "Marbhnadh Uí Carolan-Mac Cabe's Elegy on Carolan." Accessed September 26, 2016, http://irishpage.com/songs/carolan/elgycarl.htm.

Thompson, Marie, and Ian Biddle. *Sound, Music, Affect: Theorizing Sonic Experience.* New York: Bloomsbury, 2013.
Tillich, Paul. *The Courage to Be.* New Haven, CT: Yale University Press, 1952.
———. *Theology of Culture.* Edited by Robert C. Kimball. Oxford: Oxford University Press, 1959.
Tomlinson, Gary. *Monteverdi and the End of the Renaissance.* Berkeley, CA: University of California Press, 1987.
———. *Music in Renaissance Magic: Toward a Historiography of Others.* Chicago, IL: University of Chicago Press, 1993.
———. "Musical Pasts and Postmodern Musicologies: A Response to Lawrence Kramer." *Current Musicology* 53, (1993): 18–24.
Treitler, Leo. *Music and the Historical Imagination.* Cambridge, MA: Harvard University Press, 1989.
van Maas, Sander. *The Reinvention of Religious Music: Olivier Messiaen's Breakthrough toward the beyond.* New York: Fordham University Press, 2009.
———. "Introduction." In *Thresholds of Listening: Sound, Technics, Space*, edited by Sander van Maas, 1–17. New York: Fordham University Press, 2015.
Walker, Joseph C. *Historical Memoirs of the Irish Bards: Interspersed with Anecdotes of, and Occasional Observations on the Music of Ireland. Also, An Historical and Descriptive Account of the Musical Instruments of the Ancient Irish. And an Appendix Containing Several Biographical and Other Papers, with Select Irish Melodies.* London: Printed for T. Payne and Son at the Mews-Gate, and G. G. J. and J. Robinson, Pater-Noster-Row, 1786.
Walker, Steven F. "Mallarmé's Symbolist Eclogue: The 'Faune' as Pastoral." *PMLA* 93, no. 1 (1978): 106–117.
Watkins, Holly. "Schoenberg's Interior Designs." *Journal of the American Musicological Society* 61, no. 1 (2008): 123–206.
Wels, Henrik. "Late Medieval Debates on the Location of Angels after the Condemnation of 1277." In *Angels in Medieval Philosophical Inquiry: Their Function and Significance*, edited by Isabel Iribarren and Martin Lenz, 113–129. Burlington, VT: Ashgate, 2008.
Wickes, Lewis. "Schoenberg, *Erwartung* and the Reception of Psychoanalysis in Musical Circles in Vienna until 1910/11." *Studies in Music* 23, (1989): 88–106.
Williams, Sean. "Melodic Ornamentation in the Connemara Sean-nós Singing of Joe Heaney." *New Hibernia Review* 8, no. 1 (2004): 122–145.
Williams, Sean, and Lillis Ó Laoire. *Bright Star of the West: Joe Heaney, Irish Song Man.* New York: Oxford University Press, 2011.
Wolff, Christoph. *Johann Sebastian Bach: The Learned Musician.* New York: W. W. Norton, 2000.
Zbikowski, Lawrence M. "Music, Emotion, Analysis." *Music Analysis* 29, no. 1–3 (2010): 37–60.
———. "Music, Language, and Kinds of Consciousness." In *Music and Consciousness: Philosophical, Psychological, and Cultural Perspectives*, edited by David Clarke and Eric Clarke, 179–192. Oxford: Oxford University Press, 2011.
Zuidervaart, Lambert. *Adorno's Aesthetic Theory: The Redemption of Illusion.* Cambridge, MA: MIT Press, 1991.

Index

www.ingramcontent.com/pod-product-compliance
Ingram Content Group UK Ltd.
Pitfield, Milton Keynes, MK11 3LW, UK
UKHW011004180926
475701UK00004B/556
9780367231934